28. Colloquium der Gesellschaft für Biologische Chemie
21.–23. April 1977 in Mosbach/Baden

Integration and Excision of DNA Molecules

Edited by
P. H. Hofschneider and P. Starlinger

With 55 Figures

Springer-Verlag
Berlin Heidelberg New York 1978

Prof. Dr. Dr. Peter Hans Hofschneider
Max-Planck-Institut
für Biochemie
8033 Martinsried/München/FRG

Prof. Dr. Peter Starlinger
Institut für Genetik
der Universität Köln
Weyertal 121
5000 Köln 41/FRG

ISBN 3-540-08560-2 Springer-Verlag Berlin Heidelberg New York
ISBN 0-387-08560-2 Springer-Verlag New York Heidelberg Berlin

Library of Congress Cataloging in Publication Data. Gesellschaft für Biologische Chemie. Integration and excision of DNA molecules. Includes bibliographical references. 1. Genetic transformation--Congresses. 2. Transduction--Congresses. 3. Transfection--Congresses. 4. Bacteriophage--Congresses. I. Hofschneider, Peter Hans, 1929-. II. Starlinger, Peter, 1931-. III. Title. QH448.4G47. 1978. 574.8'732: 77-19175.

Offsetprinting and bookbinding: Beltz Offsetdruck, Hemsbach/Bergstr.
2131/3130-543210

Preface

The topic of this years' Mosbach Colloquium was DNA integration. We
have tried to bring together experts from different fields of research
who are studying natural processes by which DNA molecules from different sources are linked. It has been known for a long time that such
linkage occurs between the chromosomes of bacteriophages and plasmids
on the one hand and the chromosome of the bacterial host on the other.
This process has been especially well studied in bacteriophage λ. Since
it is controlled in a complicated way, we began with a lecture by
M. Ptashne on these regulatory processes. H. Nash described the integration of bacteriophage λ into the bacterial chromosome. To put this
site-specific process into perspective, G. Mosig lectured on genetic
recombination in prokaryotes in general and K. Murray described the
use of bacteriophage λ as an artificial vector for genetic engineering.

A different kind of bacteriophage integration is shown by bacteriophage
Mu, which is much less specific in its choice of an integration site
than λ. The properties of this phage were described by P. van de Putte.

While a bacteriophage is a rather independent entity, which only occa-
sionaly integrates into a bacterial chromosome, the same is not true
for transposable DNA elements, which occur only in an integrated form
within chromosomes and can jump from one region to another or even to
another chromosome carried in the same cell. Some of these chromosomes
carry genes with a recognizable function, usually resistance to an antibiotic. These transposons, as they are called, were discussed by
D.Botstein. Others are smaller and can only be recognized by the effects exerted at the site of integration. They are called insertion
sequences or IS-elements and were discussed by H. Saedler.

Do tumor viruses behave similarly to IS-elements and exert their influence by disturbing the original gene sequence? Or do they transpose genes which, when integrated at specific sites, lead to the tumor
formation? These questions and the mechanisms of viral integration for
SV 40, Adeno, Herpes and RNA Tumor Viruses were discussed by D. Nathans,
J. K. McDougal, N. Frenkel and H. E. Varmus. This was followed by a
comparison of viral and chemical carcinogenesis by Ch. Heidelberger.

The last day of the colloquium was devoted to a description of more
complicated systems, in which new linkage of DNA is made in vivo.
J.Bishop discussed the question of whether eukaryotic genomes contain
palindromic DNA sequences, which are repeatedly transposed. R. Dixon
discussed the molecular biology of nitrogen fixation, a field in which
the cooperation between bacteria and higher organisms is of importance,
and J. Schell discussed experiments on tumor formation in plants, in
which a bacterial plasmid or part of it is apparently transferred to
the plant cell, which subsequently gives rise to tumor formation.

"Human insertion elements", i.e., short communications by J. Messing,
G. Sauer, E. Fanning, and G. Bauer, contributed last-minute data relevant to the various main-topics of the symposium.

The contributions gave an interesting synopsis of experiments carried out with very different organisms. It was shown that new arrangements of DNA sequences can be produced in vivo by natural mechanisms. These are phenomena that promise,in the long run, to lead to a better understanding of the ways in which natural changes in the genome may lead to altered functions.

Apparently one can assume that the gene order in a given genome is not fixed but is modified by pathological as well as by physiological processes.

In addition, the new results on in vivo formation of new DNA combinations may lead to some arguments that will be important in the debate on cloning of new DNA combinations produced in vitro with the help of restriction endonucleases and other appropriate enzymes. Concern has been raised about this kind of research, because some scientists are afraid of the possible consequences of new combinations of DNA beginning to function in a cell in which they never existed before. These fears are raised because it is believed that such processes do not occur naturally. Should it turn out that this is not true and that new combinations of DNA are formed in vivo much more readily than has been assumed before, the fears about in vitro formation of new DNA combinations may be alleviated.

In closing, one could say that results such as the ones described here not only serve to bring together prior knowledge into a nearly perfect picture, but also opens the way to a new understanding of well-established data of classic genetics. It remains to be seen whether or not the generalized hypothesis of the "wandering genes", which has fascinated so many of us, can be verified.

We would be pleased if, upon reading this book, some colleagues show an increased interest in this field, resulting either in a critical point of view or in original experiments. For those who prefer the latter, the hopes and fears are expressed in the following verses:

> Do genes at some times leave their homes,
>   to find themselves new chromosomes?
> And, once the gene is translocated
>   does then the cell feel irritated?
> Will it perhaps react indignant
>   or - God forbid - become malignant?
> Everywhere people are wondering
>   where the genes go wandering.
> The scientist now faced with sorrow
>   may be the hero of tomorrow.
> But, then again, remember this:
>   that many a good hypothesis
> defended quietly or with a roar
>   may disappear for evermore
> and no one then must take the blame,
>   because, of course, it's all in the game!

December, 1977   Peter Hans Hofschneider
                                                 Peter Starlinger

Contents

Contributors

Baczko, K.
National Institutes of Health, Bethesda, MD, USA

Bauer, G.*
Max-Planck-Institut für Biochemie, Abt. Virusforschung, 8033 Martinsried/München, FRG

Bishop, J.M.*
Department of Microbiology, University of California Medical School, San Francisco, CA 94143, USA

Bishop, J.O.*
Department of Genetics, University of Edinburgh, West Mains Road, Edinburgh EH9 3JN, Great Britain

Bosslet, K.
Institut für Virusforschung, Deutsches Krebsforschungszentrum, 6900 Heidelberg, FRG

Chen, L.B.
Cold Spring Harbor Laboratory, Cold Spring Harbor, New York 11724, USA

Dixon, R.*
ARC Unit of Nitrogen Fixation, University of Sussex, Brighton BN1 9QJ, Great Britain

Doerfler, W.
Institut für Genetik der Universität zu Köln, Weyertal 121, 5000 Köln 41, FRG

Dunn, A.R.
Cold Spring Harbor Laboratory, Cold Spring Harbor, New York 11724, USA

Fanning, E.*
Abt. für Molekulare Genetik, Universität Konstanz, 7750 Konstanz, FRG

Frenkel, N.*
The A.J. Brandecker Laboratory, Department of Biology, University of Chicago, 920 E. 58th Street, Chicago, Ill. 60637, USA

Gallimore, P.H.
Department of Cancer Studies, University of Birmingham, Great Britain

Gellert, M.
Laboratory of Molecular Biology, NIAMDD, Bethesda, MD 20014, USA

Ghosal, D.
Institut für Biologie III, Universität Freiburg, Schänzlestraße 1, 7800 Freiburg, FRG

Giphart-Gassler, M.
Department of Molecular Genetics, State University Leiden, Leiden, Netherlands

Goosen, T.
Department of Molecular Genetics, State University Leiden, Leiden, Netherlands

---

* Speakers are marked with an asterisk.

Gronenborn, B.   Institut für Genetik der Universität zu Köln, Weyertal 121, 5000 Köln 41, FRG

Guntaka, R.   Department of Microbiology, University of California Medical School, San Francisco, CA 94143, USA

Gutai, M.W.   Department of Microbiology, Johns Hopkins University School of Medicine, Baltimore, MD 21205, USA

Heasley, S.   Department of Microbiology, University of California Medical School, San Francisco, CA 94143, USA

Heidelberger, C. *   Los Angeles County-University of Southern California, Comprehensive Cancer Center, 1721 Griffin Avenue, Los Angeles, CA 90031, USA

Hofschneider, P.H.   Max-Planck-Institut für Biochemie, Abt. Virusforschung, 8033 Martinsried/München, FRG

Hughes, S.   Department of Microbiology, University of California Medical School, San Francisco, CA 94143, USA

Kennedy, C.   ARC Unit of Nitrogen Fixation, University of Sussex, Brighton BN1 9QJ, Great Britain

Kikuchi, Y.   Laboratory of Neurochemistry, NIMH, Bethesda, MD 20014, USA

Kung, H.-J.   Department of Microbiology, University of California Medical School, San Francisco, CA 94143, USA

Leiden, J.   The A.J. Brandecker Laboratory, Department of Biology, University of Chicago, 920 E. 58th Street, Chicago, Ill. 60637, USA

McDougall, J.K. *   Cold Spring Harbor Laboratory, Cold Spring Harbor, New York 11724, USA

Meeteren, A. van   Department of Molecular Genetics, State University Leiden, Leiden, Netherlands

Messing, J. *   Max-Planck-Institut für Biochemie, 8033 Martinsried/München, FRG

Mizuuchi, K.   Laboratory of Molecular Biology, NIAMDD, Bethesda, MD 20014, USA

Montagu, M. van   State University Gent, 35, Ledeganckstraat, 9000 Gent, Belgium

Mosig, G. *   Vanderbilt University, Department of Molecular Biology, Nashville, TN 37235, USA

Müller-Hill, B.   Institut für Genetik der Universität zu Köln, Weyertal 121, 5000 Köln 41, FRG

Murray, K.*	Department of Molecular Biology, University of Edinburgh, Edinburgh EH9 3JR, Great Britain
Nash, H.A.*	Laboratory of Neurochemistry, NIMH, Bethesda, MD 20014, USA
Nathans, D.*	Department of Microbiology, Johns Hopkins University School of Medicine, Baltimore, MD 21205, USA
Phillips, C.	Department of Genetics, University of Edinburgh, West Mains Road, Edinburgh EH9 3JN, Great Britain
Ptashne, M.*	The Biological Laboratories, Harvard University, 16 Divinity Avenue, Cambridge, MA 02138, USA
Putte, P. van de*	Department of Molecular Genetics, State University Leiden, Leiden, Netherlands
Saedler, H.*	Institut für Biologie III, Universität Freiburg, Schänzlestraße 1, 7800 Freiburg, FRG
Sauer, G.*	Institut für Virusforschung, Deutsches Krebsforschungszentrum, 6900 Heidelberg, FRG
Schell, J.*	State University Gent, 35, Ledeganckstraat, 9000 Gent, Belgium
Shank, P.R.	Department of Microbiology, University of California Medical School, San Francisco, CA 94143, USA
Sutter, D.	Institut für Genetik der Universität zu Köln, Weyertal 121, 5000 Köln 41, FRG
Varmus, H.E.*	Department of Microbiology, University of California Medical School, San Francisco, CA 94143, USA
Waldeck, W.	Institut für Virusforschung, Deutsches Krebsforschungszentrum, 6900 Heidelberg, FRG
Wijffelman, C.	Department of Molecular Genetics, State University Leiden, Leiden, Netherlands

Gene Control by the Lambda Phage Repressor

M. Ptashne

The DNA of coliphage λ encodes about 50 genes. In the lysogenic state, the phage DNA molecule is integrated into the host chromosome, and most of the phage genes are turned off by the product of the phage cI gene, the so-called λ phage repressor. Upon induction of lysogens, a consequence of inactivation of repressor, the phage chromosome detaches from the host chromosome and ordinary lytic phage growth commences. The mechanism of attachment and excision of the phage chromosome is well known: A single reciprocal recombination, which normally occurs at a unique site on both the phage and host chromosome, results in integration, and excision is a reversal of this process. These remarkable recombination events are catalyzed by proteins coded by the phage and by the host, and they are discussed in detail by Howard Nash at this colloquium. The essential point to grasp here is that integration and excision do not affect gene functioning in λ. Rather, enzymes which regulate these processes are strictly controlled by regulatory proteins coded by the phage, in particular, the repressor.

The λ phage repressor binds to two regions of λ DNA, the so-called left and right operators (O_L and O_R) (see Fig. 1). Repressor bound at O_L prevents transcription of gene N whereas repressor bount at O_R prevents transcription of a gene variously called cro or tof. Partly

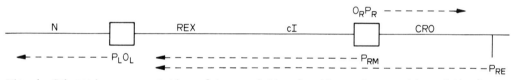

Fig. 1. Schematic representation of transcriptional patterns in a portion of the λ genome. The arrows show the directions of transcription of genes N, cro, cI, and rex. Genes cI and rex are transcribed either from the promoter P_{Rm} in lysogens or from P_{RE} after phage infection of nonlysogens. $O_L P_L$ represents the "leftward" and $O_R P_R$ the "rightward" operator promoter

because the protein of gene N is a positive regulator that is required to turn on transcription of the recombination enzymes, these enzymes are not expressed in lysogens. The repressor has evolved an elaborate mechanism for self-regulation. This ensures, among other things, that the recombination enzyme required for prophage excision is never expressed during the growth of an uninduced lysogen. Various agents including ultraviolet light and other activated carcinogens cause induction by initiating a process that results in repressor cleavage by a protease (Roberts and Roberts, 1975); we will not discuss this aspect of λ's life further here.

Work of the past several years has resulted in a rather detailed understanding of the molecular basis of gene regulation by the λ repressor. I shall consider explicitly here three examples of gene regulation involving the repressor. For a more complete discussion of these matters as well as for a list of the pertinent references, the reader is referred to Ptashne et al. (1976).

1. As mentioned above, repressor bound to O_L and O_R blocks transcription of the two adjacent genes. I shall describe the mechanisms of this negative control.

2. The cI gene is transcribed in two modes. In the lysogenic state, transcription begins near the right end of cI (near O_R) at the promoter called P_{RM} (promoter for repressor maintenance). This transcription is itself subject to both positive and negative control by repressor. Thus, the amount of repressor in lysogens is carefully regulated. I describe our current understanding of the molecular mechanisms of this autogenous control.

3. Upon infection of a nonlysogenic cell, cI transcription begins about a thousand bases to the right of cI and hence well to the right of O_R, at a promoter called P_{RE} (promoter for repressor establishment) (see Fig. 1). (Transcription beginning at P_{RE} requires the positive regulatory factors coded by the phage cII and $cIII$ genes. Once repression has been established, transcription of cII and $cIII$ is turned off by repressor, and cI is no longer transcribed from P_{RE}. The mechanism of action of cII and $cIII$ is not understood and is not considered further here.) P_{RE} directs the synthesis of five - to tenfold more repressor, per genome, than does P_{RM}, and provides the large burst of repressor necessary for the establishment of lysogeny. P_{RE} directs the synthesis of more repressor than does P_{RM} by utilizing a novel form of posttranscriptional control, as I shall describe.

Before considering these three issues, I describe our understanding of the structures of the λ operators, promoters, and repressor.

Operator Structure

The most striking aspect of the λ operators is that each contains three repressor binding sites ($O_L1,2,3$; $O_R1,2,3$). The sequences specifically recognized are 17 base pairs long and are separated by "spacers" rich in A (adenine) and T (thymine), three to seven base pairs long. The terminal binding sites O_L1 and O_R1, which are adjacent to the controlled genes N and cro, bind repressor with a higher affinity than do the remaining sites. The complete nucleotide sequences of the λ operators are shown in Figure 2. A cartoon of these sequences that emphasizes several important features is shown in Figure 3. The evidence for the preceding statements is summarized in Ptashne et al. (1976) (see also Walz et al., 1976).

Promoter Structure

A promoter is defined as a DNA sequence necessary for recognition and binding of RNA polymerase and for initiation of transcription. The promoters P_L, P_R, and P_{RM} denote, respectively, promoters for genes N, cro, and cI transcription. Figure 2 shows the regions of DNA protected from deoxyribonuclease digestion by RNA polymerase bound at these promoters. As shown first with P_R, the protected fragment is about 45 base pairs long when pancreatic deoxyribonuclease is used, and transcription begins roughly in the middle of this protected sequence. A similar relation between polymerase-protected fragments and transcription start points has been found in other cases where pancreatic deoxyribonuclease has been used. Barbara Meyer (unpublished) has recently repeated these polymerase protection experiments at P_R,

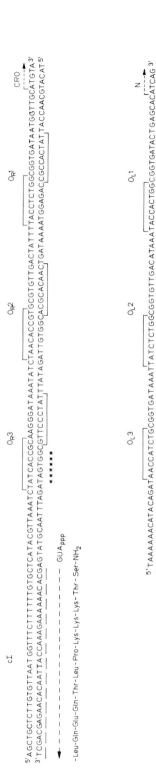

Fig. 2. DNA, RNA, and protein sequences in and around the two control regions of phage λ. The repressor binding sites $O_L1,2$, and 3 in O_L (left operator) and $O_R1,2$, and 3 in O_R (right operator) are set off in brackets. The start points of transcription of genes N, cro, and cI are indicated. Also shown are amino terminal residues of the repressor. Six bases on O_R3, presumed to code for a strong ribosome binding site for cI, are marked with an asterisk. The O_L has been reversed from its orientation in Figure 1

using λ exonuclease and the single-strand specific nuclease S1 in place of pancreatic deoxyribonuclease. Under these conditions the protected piece is roughly 65 base pairs long, and its approximate extent is indicated on the Figure. I comment below on the fact that polymerase-protected fragments overlap repressor binding sites in the operators.

Two promoter mutations have been sequenced. One, located in the spacer between O_L1 and O_L2, damages P_L. The other, located in the spacer between O_R2 and O_R3, damages P_{RM}. The former is 31 and the latter 33 base pairs from the respective start points of transcription, and each changes the sole $G \cdot C$ in a spacer to $A \cdot T$. We also know that a promoter mutation occurs in P_R within a few base pairs of the position analogous to that of the P_L mutation, but the exact base change has not been determined. As indicated in Figure 2, the RNA polymerase-protected fragment generated by pancreatic deoxyribonuclease digestion does not include the regions in which these promoter mutations occur. It is not surprising, therefore that these fragments do not bind polymerase. In contrast, the larger protected pieces obtained by λ exonuclease and nuclease S1 treatment include these regions, and these fragments bind polymerase and direct transcription. We do not know why digestion of polymerase-DNA complexes with different nucleases yields fragments of different sizes.

λ Repressor
==========

The λ repressor is an acidic protein, whose monomer has a molecular weight of about 26,000. These monomers are in concentration-dependent equilibrium with dimers and tetramers. The repressor binds tight-

4

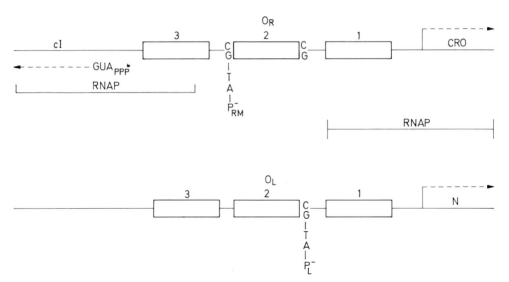

Fig. 3. Cartoon of the λ operators and a portion of genes N, cro, cI, and rex. The boxes show the positions of the 17-base pair repressor binding sites. The start point of cro, N, and P_{RM} directed cI transcription are indicated. The approximate positions of the RNA polymerase binding sites at P_L, P_R, and P_{RM}, defined as the DNA protected from deoxyribonuclease digestion by polymerase, are shown. The extents of the fragments protected (solid or dashed lines) depend on the particular deoxyribonuclease used in the protection experiment (see text). Two promoter mutations are shown

ly to DNA as an oligomer, but it is not known whether dimers or tetramers bind to the sites within the operators.

Repressor is produced in small amounts in ordinary lysogens, about 200 monomers per cell. We have constructed in vitro a recombinant DNA molecule that contains the cI gene read from two lac promoters. This recombinant is incorporated in a plasmid, and bacteria carrying this plasmid (pKB252) over-produce repressor some 50- to 100-fold. Sufficient quantities of repressor have been isolated to permit complete sequence analysis. A striking feature of the amino terminus of repressor is the strong clustering of basic residues. Although arginine and lysine constitute about 10% of the total residues, they account for 33% of the 27 amino terminal residues. It has been suggested that amino terminal residues of repressor make specific contacts with operator DNA, as has also been suggested for the lac repressor.

I now consider the three examples of gene regulation mentioned earlier.

Repressor Control of N and cro

The two terminal repressor binding sites in O_R (O_R1 and O_R2) and in O_L (O_L1 and O_L2) mediate repression of cro and N, respectively. This was deduced from the fact that mutations that render cro transcription constitutive - that is, mutations that reduce repression at that operator - have been found in O_R1 and O_R2; mutations with a similar effect on N transcription have been found in O_L1 and O_L2. At each operator, mutation of two sites has a more dramatic effect on repression than

does mutation of either site alone. RNA polymerase and repressor
binding sites overlap in each operator, and repressor excludes binding
of polymerase. Apparently repressor bound to two sites excludes polymerase more efficiently than does repressor bound to a single site
Repressor blocks transcription only if added to the template before
RNA polymerase.

Self-regulation of cI

Negative Control. Repressor bound to O_R3 turns off transcription of
cI. This was deduced as follows: RNA was transcribed in vitro from a
DNA fragment bearing P_{RM} and a portion of the cI gene, and the cI
transcript was identified by several criteria. Most important is that
this message was not produced if the template bore a mutation that
prevents transcription from P_{RM} in vivo. The sequence of the cI transcript corresponds to the DNA sequence as shown in Figures 1 and 2.
Relatively high concentrations of repressor turn off transcription of
this message in vitro and in vivo (see below). Mutations in O_R1 and
O_R2 do not drastically affect this repression as measured in vitro.
As at P_R, repressor blocks transcription only if added to the template
before RNA polymerase.

Positive Control. Repressor bound to O_R1 enhances transcription of cI.
The efficiency with which cI is transcribed in vitro can be increased
five-to tenfold by the addition of repressor. This effect requires an
intact O_R1. The mechanism of this positive effect is unknown - in
particular, we do not know the role of O_R2 - but two possibilities
are:

1.   RNA polymerase bound to P_R prevents, by steric inhibition, other
polymerase molecules from binding to P_{RM}; repressor bound to O_R1
prevents polymerase binding to P_R, but does not block access of polymerase to P_{RM}.

2.   Repressor bound to O_R1 directly enhances polymerase binding at
P_{RM}, either by providing a protein-protein contact, or by subtly al-
tering DNA structure. We note that the distance from the center of
O_R1 to the start point of transcription of cI is about the same as
that between the center of the CAP (catabolic gene activator protein)
binding site and the start point of transcription of the *lac* operon.
The CAP enhances transcription of the *lac* gene, but the mechanism is
not understood.

We favor some version of the second model, but we feel the evidence
is not conclusive.

Translational Control of cI

Why does transcription of cI initiated at P_{RE} produce more repressor
than does transcription from P_{RM}? We had anticipated a simple answer:
P_{RE} is a more efficient promoter than P_{RM}, and hence more cI transcripts are read from P_{RE}. This statement may be true, but recent
evidence indicates that a more important factor is differential translation of the messages initiated at the two promoters. Figures 2 and 3
show that the codon corresponding to the amino terminus of repressor

is found immediately adjacent to the 5' terminal AUG of the *c*I message. This is remarkable in that all messages analyzed heretofore contain leaders of variable length, preceding the AUG or GUG translational start signals. These leaders have been found to contain short sequences that are complementary to sequences at the 3' end of 16 S ribosomal RNA. It has been argued that pairing of these complementary sequences promotes binding of messages to ribosomes, and hence efficient translation. Our finding that the *c*I message transcribed from P_{RM} bears no leader suggests that it may be translated at low efficiency. In contrast, we note that beginning 12 bases to the right of the translational start point there is a six-base sequence complementary to a sequence at the 3' end of 16 S ribosomal RNA (see bases marked with an asterisk in Fig. 2). This sequence should be present in the *c*I message transcribed from P_{RE} and should function as a strong ribosome binding site. Smith et al. (1976) found that the *c*I message transcribed from P_{RE} is processed in vivo, but that the cleavage site is to the right of the proposed ribosome binding site. We conclude, therefore, that message transcribed from P_{RE} bears a strong ribosome binding site and is translated more efficiently than is message transcribed from P_{RM}.

In Summary

Let us consider the action of repressor at one operator, O_R. This operator contains three repressor binding sites designated $O_R 1$, $O_R 2$, and $O_R 3$. Because $O_R 1$ has the highest repressor affinity, at low concentrations repressor will be bound preferentially to $O_R 1$. This has the dual effect of decreasing rightward transcription of gene *cro* and of enhancing leftward transcription of the repressor gene, *c*I. At higher repressor concentrations $O_R 2$ is filled, further repressing transcription of *cro*; and at very high repressor concentrations $O_R 3$ is filled and transcription of *c*I ceases. This sequential interaction of repressor with sites within a single controlling sequence mediates negative control of a function required for lytic growth of the phage (*cro*) and autoregulates, both positively and negatively, production of repressor. Part of this same operator ($O_R 3$) codes for a sequence that apparently ensures efficient translation of *c*I, but that sequence is contained in the *c*I message only if transcription begins at one of two of the possible *c*I promoters (P_{RE}). The left operator (O_L) is similar in structure to O_R; repressor bound to $O_L 1$ and $O_L 2$ turns off transcription of another gene (*N*) required for lytic growth; the function of $O_L 3$ is not known.

The functions we have ascribed to O_L and O_R are not exhaustive. We have strong reason to believe that another repressor, the product of the *cro* gene, binds to O_L and O_R during lytic growth to turn down synthesis of *N*, *cro*, and *c*I, but how it does so remains to be seen. Moreover, the *N* protein is known to recognize some sequence in or near these operators, and to render RNA polymerase immune to the blockade found at the end of certain genes. We have not yet elucidated all the functions mediated by these remarkable regulatory sequences.

References

Ptashne, M., Backman, K., Humayun, M.Z., Jeffrey, A., Maurer, R., Meyer, B., Sauer, R.T.: Science $\underline{194}$, 156 (1976)

Roberts, J., Roberts, E.: Proc. Natl. Acad. Sci. USA $\underline{72}$, 147 (1975)

Smith, G.R., Eisen, H., Reichardt, L., Hedgpeth, J.: Proc. Natl. Acad. Sci. USA $\underline{73}$, 712 (1976)

Walz, A., Pirrotta, V., Ineichen, K.: Nature (London) $\underline{262}$, 665 (1976)

Evidence for a Complex That Coordinates Different Steps in General Genetic Recombination

G. Mosig

Illegitimate recombination, generating gross chromosomal rearrangements and gene transfer to different chromosomes or organisms, is a
major theme of this colloquium. It is not known whether this process
is in certain ways related to general genetic recombination. Both
general and illegitimate recombination are ubiquitous in nature, and
formal analysis reveals remarkable similarities in eukaryotes and
prokaryotes, suggesting that the underlying mechanisms for both kinds
of recombination are also similar. To expose differences and possible
similarities (if there are any) in these two processes, I want to discuss present knowledge and thoughts about the mechanism of general
recombination.

In contrast to its sister process, general genetic recombination is
extremely precise. It generates exchanges of *homologous* DNA segments
(and thus of genetic alleles), leaving fine structure and order of
the genetic sites on the chromosomes unaltered (Fig. 1). It involves
breakage and rejoining of DNA, accompanied by some DNA synthesis
(Meselson and Weigle, 1961; Kellenberger et al., 1961). Exchanges are
thought to occur at random positions in chromosomes, and recombination
frequencies can be used to estimate genetic distances. There are,
however, certain hot spots or hot regions of recombination that dis-
tort the congruence of genetic and physical maps (Ephrussi-Taylor,
1966; Mosig 1966; Lam et al., 1974; Henderson and Weil, 1975; Tiraby
et al., 1975; Moore and Sherman, 1975; for reviews see Stadler, 1973;
Hastings, 1975). It is not known whether these hot spots are related
to preferred sites for illegitimate recombination.

Precursors, Intermediates, and Products

In conventional genetic analyses, parental DNA is labeled with genetic
markers (mutant vs. wild-type alleles), and exchanges are deduced from
the phenotypic expression of these alleles in the progeny of genetic
crosses (cf. Fig. 1). In addition, the fate of DNA molecules under-
going genetic exchanges can be followed by biochemical means: parental
DNA is differentially labeled with density and radioisotopes, and intermediates and products are distinguished by density and sedimentation analyses and by electron microscopy. Each of these methods has
obvious advantages and disadvantages: genetic analysis can detect extremely rare events (10^{-6}) but in most cases it depends on gene expression and on production of viable recombinants. In addition, it
cannot tell whether recombinants were generated by breakage-rejoining
or by DNA synthesis and/or subsequent partial repair of heteroduplex
regions. Biophysical methods can detect nonviable intermediates, but
usually they are much less sensitive than crosses, and they rarely
permit precise localization of exchanges. Electron microscopy reveals
recombinational intermediates, but it does not distinguish parental
molecules from final recombinants. Obviously, the best understanding
of the in vivo situation comes from combining both types of analyses

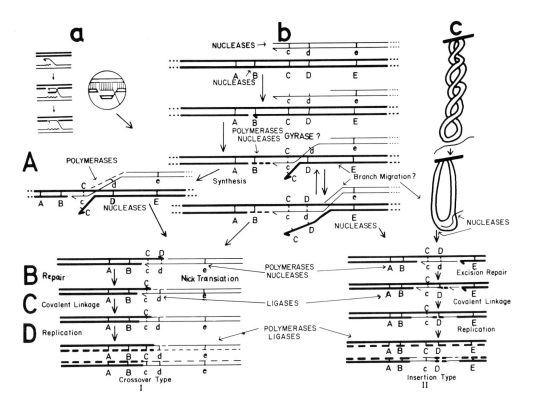

Fig. 1. Some current models of general genetic recombination. The different modes
of initiation (a - c), the different steps (A - D), and resolution into insertion-
type or crossover-type recombinants are explained in the text. Parental DNA: thick
or thin solid lines; newly synthesized DNA: broken lines. Different alleles of
different parental origin are shown as capital or lower case letters, respectively.
Both DNA strands are marked only when heteroduplexes contain different alleles; 5'
ends are marked with arrows

in the same or similar experiments. This can best be done in microorganisms, in particular in bacteriophages.

From such studies, various models of pathways of general genetic recombination have been proposed (see Fig. 1). They take into account
results of formal genetic analyses as well as known properties of DNA
and of enzymes that act on DNA. All current models are basically modifications of proposals by Whitehouse (1963) and Holliday (1964).

Branched intermediates (Broker and Lehman, 1971) are formed first
(step A): complementary single-stranded segments, derived from different parental molecules, are joined by base-pairing in heteroduplex
regions of limited length (so-called partial heterozygotes, Hershey
and Chase, 1951). This base-pairing is responsible for the high accuracy of general recombination. Molecular ends in one or in both
strands of DNA are recombinogenic (for review see Hotchkiss, 1974).
The top part of Figure 1 summarizes three plausible and mutually compatible models for initiation of such base-pairing:

1a) displacement synthesis, starting from single-strand nicks, generates single-stranded segments that can invade homologous DNA (Hotchkiss, 1974; Meselson and Radding, 1975);

1b) nucleolytic degradation from double-strand ends generates invading single-stranded segments (Broker and Lehman, 1971; Mosig et al., 1971);

1c) Holloman et al. (1975) demonstrated uptake of single-stranded DNA by unnicked supercoiled DNA in vitro. Such triple-stranded structures generate recombinants after transfection, although with rather low frequencies (Holloman and Radding, 1976).

Presumably, heteroduplex regions are short at first and are extended by single-strand branch migration (as shown) and by double-strand branch migration (not shown). Sigal and Alberts (1972) demonstrated by model-building that there is no steric hindrance to double-strand branch migration. It is not clear, however, to what extent branch migration in DNA seen in electron micrographs has occurred in vivo (Broker and Lehman, 1971) or in vitro (Lee et al., 1970).

Depending on how these intermediate structures are resolved, different types of recombinants are formed. After replication they yield two single-exchange (crossover-type) recombinants when the two new junctions occur in strands of opposite polarity (mode I, lower left panel in Fig. 1). Alternatively, such intermediates generate one double-exchange (insertion-type) recombinant and one parental type, when the two new junctions occur in the same DNA strand (mode II, lower right panel in Fig. 1). (Insertion-type recombinants should not be confused with IS elements, discussed extensively during this Colloquium!)

Note that only crossover-type recombination results in exchanges of markers outside of the heteroduplex region and thus is the main source of large-interval recombinants. In contrast, insertion-type recombination is detectable mainly in short intervals. It generates correlated double exchanges within relatively short distances but rarely recombines markers that are far apart.

Heteroduplex regions may be partially repaired (step B, Fig. 1). In many systems, most intragenic recombinants appear to be generated by heteroduplex repair (for reviews see Stadler, 1973; Hastings, 1975). Usually, heteroduplex repair alters the ratio of alleles that are recovered in the progeny. Such changes of allele ratios have been called gene conversion, long before the underlying mechanism was understood. Note that heteroduplex repair may be accomplished in two ways: it may be initiated as nick translation from an usealed junction of recombining DNA as shown in the lower left panel of Figure 1. In addition, certain mismatched base pairs may be recognized by specific enzymes and may be corrected by excision repair as shown in the lower right panel of Figure 1. (I assume that both types of repair occur in both insertion-type and crossover-type heteroduplexes, although each type is shown only once in Figure 1.) Excision repair within a heteroduplex region generates additional genetic exchanges. If it occurs within a crossover-heteroduplex, final recombinants may show multiple clustered exchanges associated with exchange of markers outside of the heteroduplex region. On the other hand, heteroduplex repair by nick translation does not generate additional genetic exchanges, but it changes allele ratios and reduces heteroduplex lengths. Thus, "gene conversion" is not necessarily associated with multiple genetic exchanges.

In step C, the exchanged DNA strands are covalently linked. Finally, DNA replication (step D) resolves unrepaired heteroduplexes to yield true recombinants.

The temporal sequence of these recombination steps is not precisely defined by experimental evidence. Presumably, in vivo these steps do not occur independently. I propose (see experiments discussed below) that most, if not all, of these steps are coordinated by binding of the responsible enzymes to a DNA binding protein, e.g., T 4 gene-*32* protein (see Fig. 5).

These recombination models are derived from studies in many different organisms (both pro- and eukaryotes) and the conclusions should have general validity. Of course, one has to keep in mind that the relative contributions to the overall outcome of various steps outlined in Figure 1 depend on the enzymatic make-up of the specific organism and on specific assay and growth conditions, e.g., whether recombination between closely or loosely linked markers is investigated and whether certain genes of DNA metabolism are altered by mutations. This has to be kept in mind when different experimental results are compared, which may at first appear contradictory.

Bacteriophage T 4 provides an excellent system for studying the mechanism of recombination, because the frequency of recombination per unit length is unusually high and because it is possible to combine both types of analyses mentioned above; often this can be done in the same experiment. All results taken together are best explained by the model shown in Figure 1.

Numerous conditional lethal mutations of T 4 have been isolated, mapped, and classified into more than 120 complementation groups (Fig. 2). The corresponding proteins have been partially characterized and account for most of the coding capacity of the 166,000 base pairs in the T 4 genome. Several genes are known to affect recombination. In T 4, molecular recombination is required for progeny production: mature DNA molecules in virions represent circular permutations of the genetic map and approximately 3% of their sequences are repeated at both ends as "terminal redundancies." After infection, crossover-type recombination has to generate DNA molecules that are longer than unit length ("concatemers") because mature DNA molecules can be packaged and cut to "head-full" size only from such concatemers (Streisinger et al., 1967). Thus, mutations that abolish all crossover-type recombination are lethal.

Genetic and molecular studies indicate that the ends of infecting T 4 DNA molecules invade each other at their ends as shown in Figure 1b (Mosig et al., 1971; Broker and Lehman, 1971), thus generating branches or loops in DNA as intermediates. Figure 3 shows one example of such looped intermediates, isolated early after infection under replication-proficient conditions by R. Dannenberg in our laboratory. Displaced single strands protrude as whiskers from many of these recombinational forks. These whiskers measure between 200 and 2000 nucleotide residues, indicating that original heteroduplex regions can be much larger than would be required for stable base-pairing. They are also larger than genetic estimates of heteroduplex lengths in mature particles (for reviews see Mosig, 1970; Broker and Doermann, 1975). This supports the conclusion (Berger and Pardoll, 1976) that there is heteroduplex repair in vivo.

Proteins Involved in T 4 Recombination

While precursors and products of recombination are rather well characterized, the precise roles of various enzymes are less well defined.

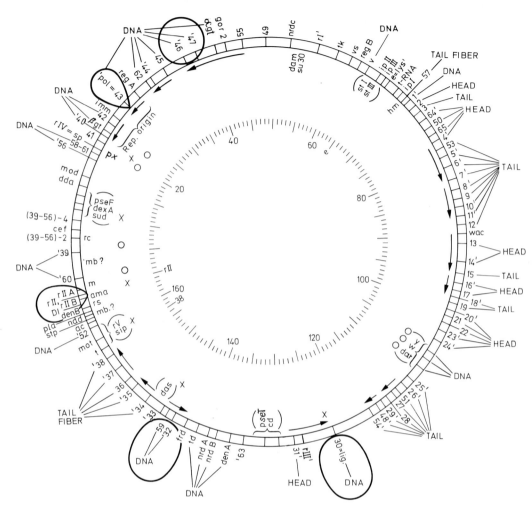

Fig. 2. Linkage map of bacteriophage T 4, modified from Mosig (1976). Genes that have been shown by our studies to interact with gene 32 are circled. Other recombination genes are marked (o). Mutations that alter expression of gene-32 mutations are marked (x). (Not all of these mutations affect gene-32 function directly)

DNA polymerases (T 4 gene *43*, *E. coli polA*), ligases (T 4 gene *30*, *E. coli lig*) and recombination nucleases (T 4 genes *46* and *47*, *E. coli rec B* and *rec C*), and DNA binding proteins (T 4 gene *32*) must participate (Figs. 2 and 6) because mutations in the corresponding genes affect recombination and because it is plausible that these enzymes affect recombination. Probably, DNA gyrase (Nash, this Colloquium) is also involved. The T 4 genes and additional recombination genes whose functions are not yet known (for recent reviews see Broker and Doermann, 1975; Miller, 1975; Cunningham and Berger, 1977) are marked in Figure 2. There are, however, many inconsistencies when one compares enzyme deficiencies in vitro and recombination deficiencies measured by different assays in vivo in many of the mutants (Bernstein, 1968; Berger et al., 1969; Allen et al., 1970; Broker, 1973; Lehman, 1974; Davis and Symonds, 1974; Naot and Shalitin, 1972; 1973; Mufti and Bernstein, 1974; Hamlett and Berger, 1975; Hosoda, 1976).

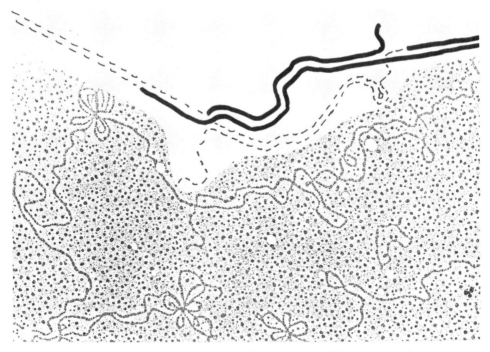

Fig. 3. A looped recombinational intermediate, isolated by R. Dannenberg early
after infection with replication proficient T 4. Two ends at different genetic
locations have become single-stranded and are invading homologous regions as postulated in Figure 1b. The distinction between replication forks and recombination
forks is described elsewhere (Mosig et al., 1975; Dannenberg and Mosig, in prep.)

Some of the apparent contradictions may simply reflect different
growth and assay conditions. For example, we have shown that the pol
A 1 mutation of *E. coli* specifically reduces T 4 crossover-type recombination while slightly enhancing insertion-type recombination or
heteroduplex repair (Mosig, 1974). On the other hand, under permissive
conditions for progeny production, certain gene-*32* (DNA binding protein)
mutations specifically reduce insertion-type recombination, but not
crossover-type recombination (Mosig et al., 1977).

It is reasonable to assume that many discrepancies between results
obtained in vivo or in vitro are due to the fact that recombination
enzymes associate with other proteins in vivo and certain mutations
affect steric or allosteric interactions (cf. Fig. 4). It is now generally accepted that proteins involved in DNA replication function as
complexes (for reviews see Kornberg, 1974; Alberts et al., 1975).
Based on our recent experiments mentioned below, we propose that the
different recombination steps outlined in Figure 1 are also catalyzed
by enzyme complexes and that certain proteins (e.g., DNA polymerase,
ligase, DNA binding protein), participate in a different manner (i.e.,
in different interactions) in recombination or replication complexes.
In such situations it is difficult to correlate biochemical interactions with the respective in vivo functions. We have used a genetic
approach, outlined in Figure 4, to help understand the biologically
important functions of such protein interactions.

14

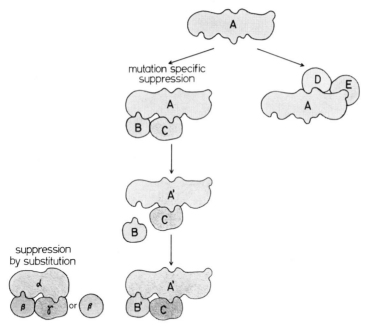

Fig. 4. A genetic approach to analyze interactions of gene products by suppressor
studies. If protein A participates in two different processes by interacting with
different cellular components, we expect that certain mutations differentially
inactivate interactions with B, thus inactivating B function, without necessarily
affecting other functions. In this situation we expect that certain site-specific
mutations in gene B partially restore interaction of B' with the altered A' protein.
Other mutations in gene B may have the opposite or no effect. Alternatively a
protein ß or a protein complex containing ß may substitute for the defective interaction of A with B. These types of interactions predict that suppression is mutation
specific (at least in one of the genes). Steric and allosteric interactions cannot
be distinguished. Genetic suppression by more indirect mechanisms (e.g., affecting
precursor pools, rates of synthesis, etc.) is expected to be gene specific but not
mutation specific

We have focused our analysis on interactions of gene-*32* protein of
phage T 4 (32 protein) because gene *32* plays a key role in DNA replication and recombination (Epstein et al., 1963; Tomizawa, 1967;
Konzinski and Felgenhauer, 1967). From the pioneering work of Alberts
and his co-workers (Alberts and Frey, 1970), it has been shown that
32-protein binds cooperatively to single-stranded DNA, that it specifically stimulates synthetic activity of T 4 DNA polymerase, facilitating
displacement synthesis (Nossal, 1974) and inhibiting nucleolytic functions of DNA polymerase (Huang and Lehman, 1972). It also stimulates
DNA synthesis in vitro by purified replication proteins (Alberts et
al., 1975). Synthesis of this protein shows autoregulation and is
induced after ultraviolet (UV) irradiation (Krisch et al., 1974; Gold
et al., 1976; Krisch and Van Houwe, 1976.

If *32*-protein participates in a different manner in DNA replication
and in recombination by interacting with DNA *and* with other proteins,
we expect that mutations that abolish binding of *32*-protein to DNA
affect most or all recombination and replication steps. On the other
hand, mutations that specifically inactivate certain protein-protein

interactions might affect only certain steps (see Fig. 4). Thus, we have asked whether certain steps in recombination as well as in DNA replication are blocked under restrictive conditions for progeny production in each of the different gene-*32* mutants shown in Figure 5.

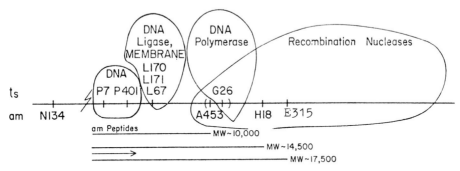

Fig. 5. Map of gene *32* mutations and interactions that are affected by these mutations. The apparent molecular weights of the *am* peptides have been determined in polyacrylamide gels (Krisch et al., 1974). Molecular weight of the wild-type protein has been reported to be 35,000 (Alberts and Frey, 1970)

To do this we have differentially labeled parental and progeny DNA with radioisotopes and with density isotopes and investigated the intracellular DNA (after infection under restrictive conditions) by density and sedimentation analyses. In addition we have measured incorporation of precursors into DNA and DNA degradation by various methods. Since host functions may partially substitute and thus mask certain defects in phage-coded functions we have also compared DNA metabolism of these gene-*32* mutants in wild-type and in mutant hosts that are defective in their own DNA metabolism. Under permissive conditions for those mutants, we have analyzed short-interval and long-interval recombination, heterozygosity, and potential repair of UV damages. All of our results (Mosig and Breschkin, 1975; Mosig and Bock, 1976; Breschkin and Mosig, 1977a, b; Mosig et al., 1977; unpublished results) show that *32*-protein functions in recombination by interacting at different sites with DNA, with recombination enzymes, and with membrane components. Under restrictive conditions, only the promoter proximal mutations *P7* and *P401* interfere with binding to DNA. All other mutations affect different interactions.

After correlating defects and map positions of the mutations (see Fig. 5), we propose that the N-terminal domain of *32*-protein is involved in binding to DNA, to membrane proteins, and to proteins that initiate DNA replication and recombination. The C-terminal domain moderates activities of recombination nucleases and thus protects DNA from excessive degradation.

All of the known gene-*32* mutations affect recombination more severely than replication. Apparently host functions can partially substitute in DNA elongation, but not in recombination (Breschkin and Mosig, 1977b).

This conclusions is further supported by analysis of secondary sup-
pressors (cf. Fig. 4). One example is the mutual suppression of one
gene-*32* mutation *(ts L 171)* and *rII* mutations (Mosig and Breschkin,
1975; Mosig et al., 1977). The *rII* proteins are membrane proteins
(Weintraub and Frankel, 1972; Ennis and Kievitt, 1972; Takacs and
Rosenbusch, 1975). The mutual suppression of *rII* and certain gene-*32*
mutations (which involves other membrane proteins and phage and host
ligase) suggests that ligation of recombining DNA, if not all recombination, occurs at the membrane. Membrane attachment would provide
topologic constraints and thus permit unwinding of linear and nicked
DNA by DNA gyrase (Nash, this Colloquium) and/or by other unwinding
proteins (Abdel-Monem et al., 1976). Similar kinds of suppressor
studies are in progress. Secondary mutations affecting expression of
certain gene-*32* mutations that have already been mapped are also
marked in Figure 2.

We suggest that by virtue of its binding to other enzymes and membrane
components, gene-*32* protein facilitates the coordinated action of
recombination enzymes on DNA in the steps outlined in Figure 1. Our
studies also indicate that specific interactions of gene-*32* protein
with T 4 ligase and T 4 DNA polymerase alter the specificity of these
enzymes in recombination and in initiation of DNA replication. In DNA
elongation, these interactions must be of a different nature, or they
are not essential for the latter process.

Recombination Proteins of *E. coli*

Properties of *E. coli* recombination genes (Fig. 6) have been summarized
by Clark (1974) (for further references see Bachmann et al., 1976).
Rec A function is most important for all general recombination as well
as for several other processes (e.g., the mutagenic SOS repair and
prophage induction e.g. see Defais et al., 1976, Mount et al. 1976).
Several treatments that cause damage to DNA induce the *rec A* functions
(for review see Witkin, 1976). It has been postulated that the *rec* A
product is or controls a DNA binding protein like the T 4 gene-*32*
protein (Grossman et al., 1975; Gudas and Pardee, 1976) and that it
is or controls a protease (Roberts and Roberts, 1975; Meyn et al.,
1977). McEntee et al. (1976) have recently purified the *rec* A protein,
but they have not reported further clues as to its function. Mutations
in genes *rec B* and *rec C* also cause recombination deficiencies. These
two genes code for part of exonuclease V. However in vitro activities
of the purified enzyme do not fully explain its role in recombination.
Lieberman and Oishi (1974) suggested that exonuclease V contains, in
addition to the *rec* BC protein, one other subunit, and Das et al.
(1976) suggested that transcription termination factor *rho* might be
this subunit. Their *rho* mutants are recombination deficient.

Two kinds of secondary suppressor mutations *(sbc A* and *sbc B)* restore
recombination proficiency in *rec B* mutants: Most interesting are the
sbc B mutations. They map in the gene for exonuclease I and require
rec A product to function in recombination. Starting with *rec B sbc B*
mutations, Horii and Clark (1973) isolated and mapped "tertiary" mutations, which again abolish recombination proficiency. It is interesting
to note that one class (*rec* F) maps near the coumermycin-resistant mutation, which affects DNA gyrase (Ryan, 1976; Gellert et al., 1976).
Another class *(rec L)* maps near the *rho* gene. A third class maps near
the original *rec B* mutation. (Two additional classes are not well
characterized.)

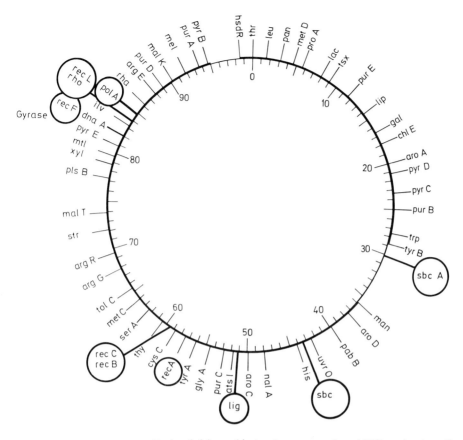

Fig. 6. Linkage map of *Escherichia coli* (Bachmann et al., 1976), showing the recombination genes mentioned here

Clark has discussed these and other mutations in terms of three recombination pathways. This view leads, however, to certain inconsistencies (Clark, 1974). It would not be surprising if these complex patterns of supression in *E. coli* also proved to reflect interactions between different *E. coli* recombination proteins.

Conclusions

How do these considerations relate to illegitimate recombination? I want to focus on two points:

1. The role of sequence homology, and
2. The functions of participating enzymes.

In most systems, general recombination between homologous DNA sequences appears to occur more frequently (per unit length DNA) than illegitimate recombination. Precise estimates are difficult because detecting illegitimate recombination usually depends on certain selection procedures.

How do homologous DNA sequences and the enzymes that mediate their
exchange recognize each other with such efficiency? How do they avoid
or correct errors? One particular source of errors is repetitive or
nearly repetitive sequences. Exchanges between such repetitive DNA
sequences at different chromosomal locations would result in duplications, deletions, and translocations. Considering the high efficiency
of general recombination, it is perhaps surprising that such rearrangements do not occur more frequently than they are found.

Are repetitive sequences too short to permit formation of stable
heteroduplexes? Is recombination between homologous or partially
homologous sequences at "wrong" locations usually prevented perhaps
by condensation of the DNA into some superstructure (Hamilton and
Pettijohn, 1976)? Is recombination between certain symmetric repetitive
sequences perhaps actively excluded?

The following consideration suggests that repetitive sequences are not
too short for stable base-pairing. All elements that are transposed
by illegitimate recombination are bordered by certain symmetric base
sequences. The potential pairing patterns between these sequences are
sufficiently stable in vitro to be revealed by hairpin loops after
denaturation and renaturation (e.g., Botstein, this Colloquium).

In addition there is considerable evidence that initial pairing of
general and of illegitimate recombination does not require perfectly
precise base-pairing. Heteroduplex regions formed during T 4 general
recombination in vivo may at first contain partially unmatched regions
that are subsequently excised (Benz and Berger, 1973).

Similarly, base sequences at the junctions of cellular and SV40 DNA
sequence, in illegitimate SV40 recombinants (Nathans, this Colloquium)
suggest that this recombination was initiated by in vivo pairing between sequences with imperfect complementarity.

Sobell (1973) has suggested that symmetric sequences are recognized
by certain proteins, thereby becoming particularly prone to initiate
site-specific as well as general recombination. While site-specific
recombination certainly occurs at such sequences (Nash, this Colloquium), there is no evidence for unusually high frequencies of general
recombination in or near such regions. To the contrary: Ely et al.
(1974) showed that certain markers within the *his* operator-promoter
region of *Salmonella* usually do not recombine. Recombination between
these markers is, however, facilitated by mutation in a (third) site
that lies between these two markers. Since this mutation affects
regulation of the *his* operon, it must affect binding of regulator
proteins. In addition, it might destroy certain symmetries within the
regulator region. Similarly, partially homologous sequences delineating
the G-loop sequence of phage Mu (see Van de Putte, this Colloquium)
are rarely, if at all, exchanged by *general* genetic recombination (D.
Kamp, personal communication). Thus, one should consider the possibility that binding of certain proteins to certain symmetric sequences
prevents binding of proteins that are specific for general recombination, while still allowing or promoting binding of proteins that are
required for illegitimate recombination.

We conclude from our analysis of T 4 recombination that several proteins that participate in different processes of DNA metabolism, e.g.,
DNA replication and recombination, are modified and channeled into
recombination pathways by associating at different specific sites with
a DNA-binding protein. I want to emphasize two implications for the
major theme of this meeting:

1. It is generally believed that illegitimate and general recombination are promoted by different enzyme systems, because *E. coli rec A* mutations do not interfere with illegitimate recombination. This does not exclude the possibility, however, that both of these processes share common components. Perhaps the *rec A* protein channels other, common enzymes mainly into general recombination. It is also possible that the available *rec A* mutations do not inactivate all *rec A* functions or that other functions can substitute for *rec A* in illegitimate but not in general recombination. It is intriguing to note that mutations that affect deletion formation related to IS integration (Saedler, this Colloquium), map near *rec B/C*.

2. The integration of viral DNA sequences into host DNA by recombination may involve modifications of host enzymes by virus-coded DNA binding proteins similar to the modifications by *32*-protein that we have found.

Acknowledgements. I thank Drs. Ruth Ehring and Peter Starlinger for many stimulating discussions, J. Straub for hospitality, and the Alexander von Humboldt Stiftung for generous support during my stay in Köln. Work in our laboratory in Nashville has been supported by NIH grant GM 13221.

References

Abdel-Monem, M., Dürwald, H., Hoffmann-Berling, H.: Eur. J. Biochem. 65, 441-449 (1976)

Alberts, B., Frey, L.: Nature 227, 1313-1318 (1970)

Alberts, B., Morris, C.F., Mace, D., Sinha, N., Bittner, M., Moran, L.: In: DNA Synthesis and its Regulation. Goulian, M., Hannwalt, P. (eds.). Menlo Park, Calif. Benjamin Inc.: 1975, pp. 241-295

Allen, E.F., Albrecht, I., Drake, J.W.: Genetics 65, 187-200 (1970)

Bachmann, B.J., Low, K.B., Taylor A.L.: Bacteriol. Rev. 40, 116-167 (1976)

Benz, W.C., Berger, H.: Genetics 73, 1-11 (1973)

Berger, H., Pardoll, D.: J. Virol. 20, 441-445 (1976)

Berger, H., Warren, A.J., Fry, K.E.: J. Virol. 3, 171-175 (1969)

Bernstein, H.: Cold Spring Harbor Symp. Quant. Biol. 33, 325-331 (1968)

Breschkin, A.M., Mosig, G.: J. Mol. Biol. 112, 279-294 (1977a)

Breschkin, A.M., Mosig, G.: J. Mol. Biol. 112, 295-308 (1977b)

Broker, T.R.: J. Mol. Biol 81, 1-16 (1973)

Broker, T.R., Doermann, A.H.: Ann. Rev. Genet. 9, 213-244 (1975)

Broker, T.R., Lehman, I.R.: J. Mol. Biol. 60, 131-149 (1971)

Clark, A.J.: Genetics 78, 259-271 (1974)

Cunningham, R.P., Berger, H.: Virology 80, 67-82 (1977)

Das, A., Court, D., Adhya, S.: Proc. Natl. Acad. Sci. USA 73, 1959-1963 (1976)

Davis, K.J., Symonds, N.: Mol. Gen. Genet. 132, 173-180 (1974)

Defais, M., Caillet-Fauquet, P., Fox, M.S., Radman, M.: Mol. Gen. Genet. 148, 125-130 (1976)

Ely, B., Fankhauser, D.B., Hartman, P.E.: Genetics 78, 607-631 (1974)

Ennis, H.L., Kievitt, K.D.: Proc. Natl. Acad. Sci. USA 70, 1468-1472 (1973)

Ephrussi-Taylor, H.: Genetics 54, 211-222 (1966)

Epstein, R.H., Bolle, A., Steinberg, C.M., Kellenberger, E., Boy De La Tour, E., Chevalley, R., Edgar, R.S., Susman, M., Denhardt, G.H., Lielaulis, A.: Cold Spring Harbor Symp. Quant. Biol. 28, 375-394 (1963)

Gellert, M., O Dea, M.H., Itoh, T., Tomizawa, J.-I.: Proc. Natl. Acad. Sci. USA 73, 4474-4478 (1976)

Gold, L., O'Farrell, P.Z., Russel, M.: J. Biol. Chem. 251, 7251-7262 (1976)

Grossman, L., Braun, A., Feldberg, R., Mahler, I.: Annu. Rev. Biochem. 44, 19-43 (1975)

Gudas, L.J., Pardee, A.B.: J. Mol. Biol. 101, 459-477 (1976)
Hamilton, S., Pettijohn, D.E.: J. Virol. 19, 1012-1027 (1976)
Hamlett, N., Berger, H.: Virology 63, 539-567 (1975)
Hastings, P.J.: Annu. Rev. Genet. 9, 129-144 (1975)
Henderson, D., Weil, J.: Genetics 79, 143-174 (1975)
Hershey, A.D., Chase, M.: Cold Spring Harbor Symp. Quant. Biol. 21, 91-101 (1951)
Holliday, R.: Genet. Res. 5, 282-304 (1964)
Holloman, W.K., Radding, C.: Proc. Natl. Acad. Sci. USA 73, 3910-3914 (1976)
Horii, Z.I., Clark, A.J.: J. Mol. Biol. 80, 327-344 (1973)
Hosoda, J.: J. Mol. Biol. 106, 277-284 (1976)
Hotchkiss, R.D.: Annu. Rev. Microbiol. 28, 445-468 (1974)
Huang, W.M., Lehman, I.R.: J. Biol Chem. 247, 3139-3146 (1972)
Kellenberger, G., Zichichi, M.L., Weigle, J.J.: Proc. Natl. Acad. Sci. USA 47,
    869-878 (1961)
Kornberg A.: DNA Synthesis. San Francisco: W.H. Freeman and Co. 1974
Kozinski, A., Felgenhauer, Z.: J. Virol. 1, 1193-1202 (1967)
Krisch, H.M., Bolle, A., Epstein, R.H.: J. Mol. Biol. 88, 89-104 (1974)
Krisch, H.M., Van Houwe, G.: J. Mol. Biol. 108, 67-81 (1976)
Lam, S.T., Stahl, M.M., McMilin, K.D., Stahl, F.W.: Genetics 77, 425-433 (1974)
Lee, C.S., Davis, R.W., Davidson, N.: J. Mol. Biol. 48, 1-22 (1970)
Lehman, I.R.: Science 186, 790-797 (1974)
Lieberman, R.P., Oishi, M.: Proc. Natl. Acad. Sci. USA 71, 4816-4829 (1974)
McEntee, K., Hesse, J.E., Epstein, W.: Proc. Natl. Acad. Sci. USA 73, 3979-3983
    (1976)
Meselson, M., Weigle, J.J.: Proc. Natl. Acad. Sci. USA 47, 857-868 (1961)
Meselson, M.S., Radding, C.M.: Proc. Natl. Acad. Sci. USA 72, 358-361 (1975)
Meyn, M.S., Rossman, T., Troll, W.: Proc. Natl. Acad. Sci. USA 74, 1151-1156 (1977)
Miller, R.C., Jr.: Annu. Rev. Microbiol. 29, 355-376 (1975)
Moore, C.W., Sherman, F.: Genetics 79, 397-419 (1975)
Mosig, G.: Proc. Natl. Acad. Sci. USA 56, 1177-1183 (1966)
Mosig, G.: In: Advances in Genetics. New York: Academic Press, 1970, Vol. 15, pp.
    1-53
Mosig, G.: In: Mechanisms in Recombination. Grell, R.F. (ed.) New York: Plenum
    Publishing Corp., 1974, pp. 29-39
Mosig, G.: In: Handbook of Biochemistry and Molecular Biology, 3rd ed. Fasman, G.
    (ed.). CRC Press, Cleveland, Ohio 1976, pp. 664-676
Mosig, G., Berquist, W., Bock, S.: Genetics 86, 5-23 (1977)
Mosig, G., Bock, S.: J. Virol. 17, 756-761 (1976)
Mosig, G., Breschkin, A.: Proc. Natl. Acad. Sci. USA 72, 1226-1230 (1975)
Mosig, G., Dannenberg, R., Breschkin, A.: Indian J. Microbiol. 15, 145-160 (1975)
Mosig, G., Ehring, R., Schliewen, W., Bock, S.: Mol. Gen. Genet. 113, 51-91 (1971)
Mount, D.W., Kosel, C.K., Walker, A.: Mol. Gen. Genet. 146, 37-41 (1976)
Mufti, S., Bernstein, H.: J. Virol. 14, 860-871 (1974)
Naot, Y., Shalitin, C.: J. Virol. 10, 858-865 (1972)
Naot, Y., Shalitin, C.: J. Virol. 11, 862-871 (1973)
Nossal, N.C.: J. Biol. Chem. 249, 5668-5676 (1974)
Roberts, J.W., Roberts, C.W.: Proc. Natl. Acad. Sci. USA 72, 147-151 (1975)
Ryan, M.J.: Biochemistry 15, 3769-3777 (1976)
Sigal, N., Alberts, B.: J. Mol. Biol. 71, 789-793 (1972)
Sobell, H.M.: Adv. Genet. 17, 411-490 (1973)
Stadler, D.R.: Annu. Rev. Genet. 7, 113-127 (1973)
Streisinger, G., Emrich, J., Stahl, M.M.: Proc. Natl. Acad. Sci. USA 57, 292-295
    (1967)
Takacs, B.J., Rosenbusch, J.P.: J. Biol. Chem. 250, 2339-2350 (1975)
Tiraby, G., Fox, M.S., Bernheimer, H.: J. Bacteriol. 121, 608-618 (1975)
Tomizawa, J.-I.: J. Cell. Physiol. 70, suppl. 1, 201-214 (1967)
Weintraub, S.B., Frankel, F.R.: J. Mol. Biol. 70, 589-615 (1972)
Whitehouse, H.L.L.: Nature (London) 199, 1034-1040 (1963)
Witkin, E.W.: Bacteriol. Rev. 40, 869-907 (1976)

Integrative Recombination of Bacteriophage Lambda: Genetics and Biochemistry

H. A. Nash, Y. Kikuchi, K. Mizuuchi, and M. Gellert

Introduction

Integrative recombination is the process by which the genome of bacteriophage λ is linearly inserted into the genome of its *Escherichia coli* bacterial host. This recombination has been, and continues to be, intensively studied from genetic, physiologic, and biochemical points of view. As a result there has been much progress in our understanding of the mechanism of this recombination, and this paper details several of these recent developments.

Knowledge of some basic facts concerning the genetics of lambda integration is fundamental to appreciation of recent biochemical advances. The following capsule summary may be supplemented with recent reviews of the field by Campbell (1976) and Weisberg et al. (1977). More than 15 years ago Campbell suggested that λ integration proceeds by a reciprocal recombination between circular forms of the λ chromosome and the *E. coli* genome. Work by several groups of molecular biologists in the intervening years has totally verified this proposal, and, in addition, has characterized three interacting components of this recombination - the *int* gene, the phage attachmend site *att*P, and the bacterial attachment site *att*B (Fig. 1). The *int* gene is the only phage gene known that is required for λ integrative recombination; it maps adjacent to the phage attachment site and appears to code for a polypeptide product of approximate molecular weight equal to 42,000 daltons. The sites at which integrative recombination occurs, *att*P and *att*B, are genetically distinguishable from one another (hence the separate names). These sites are small, comprising less than one hundred base pairs, but they are recognized with high accuracy and efficiency, since integration of the virus is both a precise and frequent event. Moreover they are recognized asymmetrically, since the integrated viral DNA is always present in a unique orientation, with the *int* gene close to the bacterial galactose operon. No physical or genetic homology can be detected between *att*P and *att*B, indicating that integrative recombination does not require pairing of long stretches of identical DNA, in contrast to the general recombination systems (*rec*- and *red*-promoted) of *E. coli* and phage λ.

Int-promoted recombination is not limited to interaction between phage and host chromosomes. For example, recombination between *att*P and *att*B occurs when both sites are carried on the same transducing phage DNA chromosome (Fig. 2). In this arrangement, the two sites located on the same DNA molecule recombine with each other preferentially, and the overall efficiency of recombination is greatly enhanced (Nash, 1974). This enhanced recombination presumably reflects the inability of the attachment sites to diffuse away from each other because of their connection to one another by their common DNA backbone. Such intramolecular site-specific recombination has proved to be a great convenience in physiologic studies and furthermore has been essential for the production of recombinant DNA in cell-free extracts.

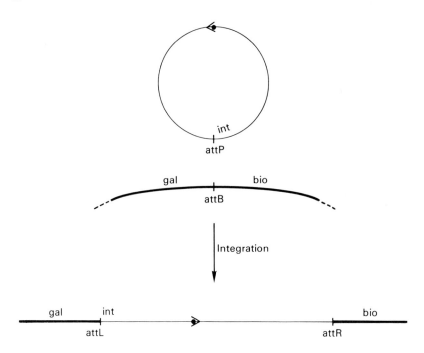

Fig. 1. The Campbell model of λ integration. Bacterial and phage DNA are shown as heavy and light lines, respectively. The ball and arrowhead indicate the cohesive termini of the mature phage DNA that are sealed after infection by *E. coli* DNA ligase

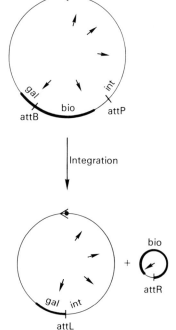

Fig. 2. λ*att*B-*att*P and its recombinant products. The virus whose genome is shown at the top of the Figure was constructed by in vivo illegitimate recombination between λ and a λ-transducing phage carrying *att*B (Nash, 1974). The products of integrative recombination of this DNA (shown at bottom) contain 87% and 13% of the parental DNA. The arrows point to the approximate location of the targets for restriction endonuclease EcoRI. The two restriction fragments of parental DNA that abut against the restriction target in the bio region are missing in the recombinant DNA. They are replaced by a fragment (called Rb) that includes *att*L and a linear fragment (Ra) produced by cleavage of the 13% *att*R-containing circle

Integrative Recombination in vitro

In 1975 a system was described for the production of integrative recombinant DNA by cell-free extracts (Nash, 1975). The substrate was
the DNA of λattB-attP phage (Fig. 2). Conversion of this substrate to
phage DNA molecules that are deleted for the block of biotin genes
was assayed by analysis of the mature phage particles produced when
the DNA taken from in vitro reactions was transfected in spheroplasts.
Recombination in vitro was shown to resemble recombination in infected
cells in several ways. The recombination takes place only at the attachment sites of the parental DNA and produces the appropriate recombinant prophage attachment site. In addition, there is an absolute
requirement for cell-free extracts which are derived from cells containing an active int gene. Recombination in vitro is an efficient
process - under optimal conditions more than 50% of the transfected
DNA molecules give rise to phage particles of recombinant genotype.
An initial biochemical characterization of the components of the reaction showed that both phage- and host-coded enzymes were required
for the reaction and that both spermidine and ATP were required as
organic cofactors.

Molecular Forms of Substrate DNA

Comparison of the efficiency of different forms of λ attB-attP DNA as
substrate for recombination in vitro reveals a pronounced bias in
favor of covalently closed circular DNA. Linear DNA and circles with
naturally occuring interruptions or nicks introduced by treatment with
X-ray or nuclease yield no detectable recombination in vitro. Under
the same conditions covalently closed circles are almost completely
converted from the substrate to recombinant form (Mizuuchi and Nash,
1976). The comparison of these different molecular forms in vitro has
been greatly obviated by the development of a new assay for recombination in vitro. This assay detects the recombination-dependent rear-
rangement of the substrate DNA by comparison of the restriction endonuclease digest of parental and recombinant DNA (Fig. 2). This relatively direct and physical method has been used to show that a complete
cycle of breakage and reunion occurs during recombination in vitro and
that the various molecular forms of the substrate DNA persist un-
degraded during the reaction, ruling out a trivial explanation for
their different efficiencies (Mizuuchi and Nash, 1976).

The preference for closed circle DNA as substrate for recombination
in vitro is a reflection of its ability to become supercoiled. Indeed,
in reaction mixtures that lack ATP, negatively supertwisted DNA is the
only effective substrate (Fig. 3, lanes 2 and 4). The mechanism underlying the preference for supertwisted DNA is not yet known. An attractive possibility is that the energy of supercoiling assists in the
binding of some required recombination component. A related alternative is that the torsional strain of the suptertwisted DNA substrate
promotes the melting of the double helix in some critical region of
the substrate, such as the attachment sites. Further study of the
interaction between substrate DNA and purified components of in vitro
recombination mixtures should clarify this issue.

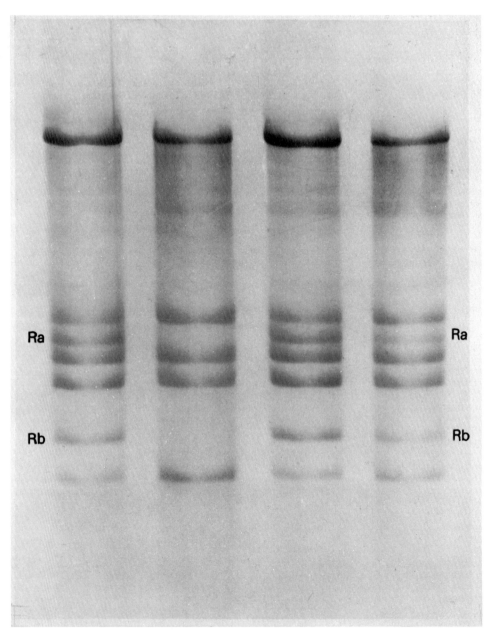

Fig. 3. Recombination in vitro with closed circular λattB-attP DNA substrates. Starting at the left, lanes 1 and 2 display restriction digests of reactions with relaxed closed circular substrate; lanes 3 and 4 are assays with supertwisted substrate. ATP was omitted from the reactions assayed in lanes 2 and 4; otherwise recombination was performed as described in Mizuuchi and Nash (1976)

DNA Gyrase

In the presence of ATP, closed circular DNA without supertwists is almost as efficient a substrate for integrative recombination as supertwisted substrate (Fig. 3, lanes 1 and 3). This observation indicates that the extracts used for recombination in vitro contain an activity that carries out ATP-dependent conversion of relaxed closed circular DNA into supertwisted DNA. Indeed, column chromatography of the host-coded extract separates a component that is required, together with ATP, for recombination using closed circular relaxed substrate but is inessential for recombination with supertwisted DNA substrate. This activity has been named DNA gyrase and has been further purified using a more direct assay for the conversion of relaxed to supertwisted DNA (Gellert et al., 1976a). The enzyme has cofactor requirements for ATP and Mg; spermidine stimulates its activity but is not an absolute requirement. In glycerol gradients, DNA gyrase activity cosediments with globular proteins of molecular weight 140,000.

DNA gyrase is an essential cellular enzyme. The antibiotics novobiocin and coumermycin Al are lethal to bacteria as a result of their ability to inhibit this enzyme. DNA gyrase activity is inhibited by these antitiotics when the enzyme is extracted from wild-type, antibiotic-sensitive cells, and gyrase activity is resistant to these antibiotics when the enzyme is extracted from bacteria that carry a single, antibiotic resistance mutation (Gellert et al., 1976b). Recently, integrative recombination has been investigated in whole cells treated with coumermycin Al. At those concentrations of antibiotic that inhibit the intracellular supercoiling of lambda DNA in infected cells, integrative recombination is reduced more than 20-fold (H. Nash and Y. Kikuchi, unpublished observations). This strongly suggests that, in vivo as well as in vitro, supertwisted DNA is the substrate for integrative recombination. Furthermore, in antibiotic-treated cells, recombination between *att*P and *att*B carried on separate phage chromosomes is also abolished. Therefore, supertwisted DNA appears to be required for both intermolecular, as well as intramolecular recombination.

What is the evolutionary significance of the apparent requirement for supertwisted DNA in integration? It may simply be that DNA circles are maintained as supercoils in connection with other intracellular processes and that enzymatic mechanisms that have evolved to operate on intracellular circular DNA have no choice but to recognize and utilize supercoils. An alternative, more homeostatic mechanism may be proposed. The requirement for supertwists could be a mechanism to insure that integrative recombination is limited to circles. Reciprocal recombination between a circular bacterial chromosome and a linear viral DNA would introduce a potentially lethal discontinuity into the host genome. Supercoiling is a process that is available to circular DNA but not to linear DNA. Perhaps the recombination mechanism has evolved to use supercoils as a physical indication of the circularity of a potential substrate. Comparative studies of excisive recombination of lambda DNA or integrative recombination with other viruses should help in assessing the proposal.

Host-coded Genes and Enzymes

A reaction mixture that contains all the known cofactors, supertwisted substrate DNA, and partially purified *int* gene product carries out

little or no recombination in vitro. Such a mixture must be supplemented by additional protein components, which can be extracted from
nonlysogenic bacteria. What is the nature of these components? Two
enzymes from *E. coli*, DNA ligase and DNA gyrase, known to be involved
in preparation of the substrate, have been ruled out as playing a
further role in recombination. In complete reaction mixtures with
closed circle substrate, recombination in vitro proceeds efficiently
even if ligase activity is blocked by addition of a specific inhibitor
(nicotinamide mononucleotide) or by mutation (lig ts7). Similarly,
efficient recombination occurs with supertwisted substrate even if DNA
gyrase activity is removed by addition of a specific inhibitor (coumermycin Al) or by chromatography. Recently, two proteins that are
encoded by the bacterial host and are required for integrative recombination have been identified by mutation. An initial mutant defective in lambda integration was isolated serendipitously (Williams et
al., 1977); many more mutants that affect the same cistron, called
*him*A, have now been isolated following a rational selection procedure
(Miller and Friedman, 1977). In addition, mutations at a second locus,
called *hip*, have been isolated following a different selection procedure (A. Kikuchi and R. Weisberg, personal communication). The map
location of these loci is shown in Figure 4.

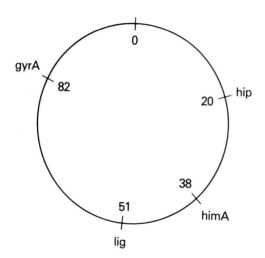

Fig. 4. Genetic map of *E. coli*
showing position of loci involved
in integrative recombination. The
coordinates are the recalibrated
units of Bachmann et al. (1976).
E. coli ligase has been mapped by
Gottesman et al. (1973). The locus
for DNA gyrase is taken from the
mapping of coumermycin Al-resistant
alleles (Ryan, 1976). The location
of *him* and *hip* is deduced by their
cotransduction with *aro*D and *aro*A,
respectively (H. Miller and D.
Friedman, personal communication,
A. Kikuchi and R. Weisberg, personal
communication)

Typical mutants of both classes have been analyzed for recombination
in vitro and display similar phenotypes. The mutants are unaffected
in their ability to produce functional *int* gene product, indicating
that their defect in integration is not *via* regulation of phage gene
product synthesis. Moreover, cell-free extracts from mutant bacteria
are defective in recombination in vitro even when supertwisted substrate and an appropriate source of *int* gene product and cofactors
are supplied. That is to say, the mutants appear to be defective in
the host-coded components required for integrative recombination.
Extracts of mutant bacteria do not inhibit extracts of wild-type
bacteria nor do such extracts damage substrate DNA significantly more
than extracts of wild-type bacteria. Recently, it has been found that
mutant extracts of *him*A bacteria complement extracts from *hip* bacteria
in vitro, indicating that each locus codes for a separate polypeptide
that is involved in integrative recombination (H. Nash, unpublished
observation).

Acknowledgements. We thank David Friedman, Aki Kikuchi, Harvey Miller, and Robert Weisberg for communicating results to us prior to publication. We also thank Dr. P. Botchan for suggesting possibilities for the biologic relevance of a superhelical substrate.

References

Bachmann, B.J., Low, K.B., Taylor, A.L.: Bacteriol. Rev. $\underline{40}$, 116-167 (1976)

Campbell, A.M.: Sci. Am. $\underline{235}$, 103-113 (1976)

Gellert, M., Mizuuchi, K., O'Dea, M.H., Nash, H.A.: Proc. Natl. Acad. Sci. USA $\underline{73}$, 3872-3876 (1976a)

Gellert, M., O'Dea, M.H., Itoh, T., Tomizawa, J.: Proc. Natl. Acad. Sci. USA $\underline{73}$, 4474-4478 (1976b)

Gottesman, M.M., Hicks, M.L., Gellert, M.: J. Mol. Biol. $\underline{77}$, 531-547 (1973)

Miller, H.I., Friedman, D.I.: In: Plasmids, DNA Insertion Elements and Episomes. Shapiro, J., Bukhari, A., Adhya, S. (eds.) Cold Spring Harbor Laboratory: N.Y., 1977, pp. 349-356

Mizuuchi, K., Nash, H.A.: Proc. Natl. Acad. Sci. USA $\underline{73}$, 3524-3528 (1976)

Nash, H.A.: Virology $\underline{57}$, 207-216 (1974)

Nash, H.A.: Proc. Natl. Acad. Sci. USA $\underline{72}$, 1072-1076 (1975)

Ryan, M.J.: Biochemistry $\underline{15}$, 3769-3777 (1976)

Weisberg, R.A., Gottesman, S., Gottesman, M.E.: In: Comprehensive Virology. Frankel-Conrat, H., Wagner, R. (eds.) Plenum Press: N.Y., 1977, Vol.III, pp. 197-258

Williams, J.G.K., Wulff, D.L., Nash, H.A.: In: Plasmids, DNA Insertion Elements and Episomes. Shapiro, J., Bukhari, A., Adhya, S. (eds.) Cold Spring Harbor Laboratory, N.Y., 1977, pp. 357-361

Applications of Bacteriophage λ in Recombinant DNA Research

K. Murray

The extensive background of biochemical genetics of bacteriophage
lambda makes the virus particularly useful in genetic engineering experiments. Genetic and biochemical approaches have been used to con-
struct phages with genomes such that they can function as receptors
for fragments of DNA. Recombinant DNA molecules made in vitro can then
be recovered by transfection of competent *Escherichia coli* cultures
or by packaging into phage particles in vitro. Several procedures are
available for screening or selection of the recombinant phages and in
the case of recombinants containing segments of eukaryotic DNA, hybridization methods are applicable to the screening of individual phage
plaques for the presence of sequences complementary to that of a
suitable radioactive probe, such as a labelled mRNA preparation. The
regulatory systems of lambda can be exploited in a number of ways to
optimise the expression of inserted DNA fragments and with prokaryotic
DNA the yield of gene products may thus be enhanced. The recombinant
phages can, of course, be further manipulated themselves to improve
yields of DNA or of gene products, or to facilitate the analysis of
cloned DNA sequences. Separation of the two strands of lambda DNA is
useful in the latter context and in ascertaining the orientation, and
hence the direction, of transcription of an inserted DNA sequence with
respect to that of the phage chromosome.

Single-Stranded Filamentous DNA Phage as a Carrier for in vitro Recombined DNA

J. Messing, B. Gronenborn, B. Müller-Hill, and P. H. Hofschneider

Introduction

In contrast to bacteriophage λ the group of single-stranded DNA phages may offer some special features in recombinant DNA research. Owing to the life cycle of filamentous phages (Fig. 1), single-stranded and

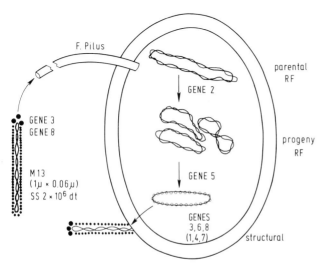

SS = SINGLE STRANDED DNA
RF = REPLICATIVE FORM

Fig. 1. Life cycle of filamentous phages. Viral DNA is a single-stranded covalently closed DNA with a molecular weight of 2×10^6 daltons, which is packaged in a filamentous protein envelope. After infection of an F^+ strain of *Escherichia coli*, phage DNA is converted into a double-stranded supercoiled replicative form and amplified to 200 - 300 copies per cell. Late in infection single-stranded viral DNA is produced and after maturation extruded from the cell without lysis. For further details see Marvin and Hohn (1969)

double-stranded DNA can be isolated at the same time. Thus integrated foreign DNA into the phage genome can easily be sequenced following the procedure applied for ØX 174 by Sanger and Coulson (1975). Since the phage DNA is packaged into a filamentous protein envelope, the size of inserted DNA is not restricted to one unit length of phage genome. Host cells are productive in viral synthesis without loss of viability. About 200 - 300 copies of replicative double-stranded supercoiled molecules can accumulate per cell (Marvin and Hohn, 1969). Infected cells continue to produce viral products, but neither lyse nor stop growing. Growth is slightly retarded compared to uninfected cells and enables the formation of "turbid plaques" within a bacterial lawn that can be used to detect infected cells.

The Search for a Suitable Recombination Site in Filamentous Phage DNA

The λ genome is known to contain genes that are not essential for the replication of the phage. As described in the preceding paper the substitution of nonessential restriction fragments by a piece of foreign DNA leads to the construction of recombinant phage DNA (Murray and Murray, 1974; Rambach and Tiollais, 1974; Thomas et al., 1974). In contrast to λ the eight known and mapped genes of filamentous phages (v.d. Hondel et al., 1976; Edens et al., 1976) are essential, and inactivation of any one of them affects the viability of the phages.

However, assuming at least one nonessential stretch of DNA a suitable site of recombination can be expected between these genes. By cutting the phage DNA statistically at several sites and by inserting a DNA fragment of known function, such a site could be identified.

Construction of the Hybrid Phage M13 mp1

M13 wild-type RF has been digested with restriction endonuclease BsuI at room temperature to a limited extent. The reaction products were separated by gel electrophoresis and the linearized RF III molecules were sliced out and further purified. There are ten BsuI targets on the M13 RF DNA, which have been ordered on a circular map by v.d. Hondel et al. (1976) as shown outside of the circle in Figure 2. The randomly cleaved RF molecules were joined to a HindII fragment of the lac operon DNA (Fig. 3) by blunt-end ligation. Successful insertion of this piece of *lac* DNA can be tested by a color reaction of β-galactosidase. If *lac* DNA is cleaved by HindII endonuclease a fragment of about 800 base pairs (Fig. 3) can be retained on nitrocellulose filters by the *lac* repressor binding assay (Landy et al., 1974). This fragment comprises the *lac* regulatory region and the information for the first 145 amino acids of β-galactosidase (Gilbert, W. personal communication). This aminoterminal fragment of β-galactosidase is able to complement the promoter proximal M15 deletion of the Z gene in an appropriate host (Landy et al., 1974).

The reactants of the ligation were subjected to transfection of competent cells of a strain suitable for the described β-galactosidase complementation assay. In the presence of IPTG, the allosteric effector of the *lac* repressor, and Xgal, a color indicator for active β-galactosidase (Miller, 1972), two blue plaques were found among several hundred colorless ones, indicating complementation. Hybrid phages termed M13 mp1 were isolated and purified.

Mapping of the Insertion Point of Recombined lac DNA

The insertion of the *lac* DNA has been mapped at the BsuI site at 0.083 map units by heteroduplex analysis. This site is located in a region between the origin of replication and the promoter controlling the expression of gene II (Fig. 2). These results can be confirmed by restriction endonuclease cleavage analysis. Furthermore the orientation of the inserted *lac* DNA as analyzed by the cleavage pattern of BsuI and HpaII reveals that transcription from the *lac* promoter occurs in the same direction as all transcription of viral genes (Fig. 2).

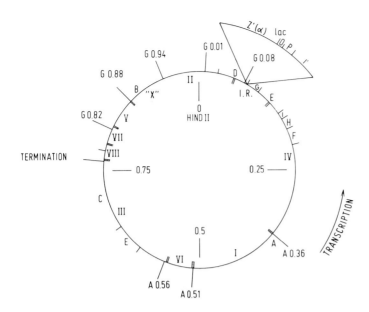

Fig. 2. Physical and genetic map of M13 wild-type and M13 mp1. The physical and genetic map of M13 wild-type is represented by a circle with the positions of viral functions inside of the circle and the order of the BsuI fragments outside of the circle as worked out by v.d. Hondel et al. (1976). The location of the three A and the five G promoters has been shown by Edens et al. (1976). The inserted *lac* fragment is represented by an outer circle segment on the same scale of map units as the inner one.

I': part of the *lac* repressor gene; p: *lac* promoter;
o: *lac* operator; Z' (α): α-region of β-galactosidase

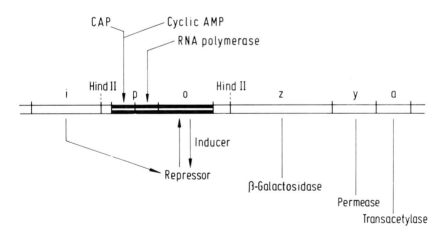

Fig. 3. Map of the lac operon. The promoter-operator region is somewhat expanded, the flanking HindII cuts are as indicated

Concluding Remarks

The inserted *lac* fragment improves the carrier properties of the phage in a number of ways. A restriction endonuclease site that would be unique for M13 mp1, as for instance that generated by EcoRI, may be introduced in the *lac* Z region as a receptor site for DNA fragments. The complementation assay may be used as a screening method for cloned DNA sequences. The regulatory system of *lac* DNA can be employed for the expression of inserted DNA fragments and the yield of gene products can be enhanced owing to the strong *lac* promoter and the gene dosage of the multiple copies of intracellular M13 mp1 RF of infected cells.

Acknowledgement. This work was supported by the Deutsche Forschungsgemeinschaft.

References

Edens, L., van Wezenbeck, P., Konigs, R.N.H., Schoenmakers, J.G.G.: Eur. J. Biochem. 70, 577 (1976)

v.d. Hondel, C.A., Pennings, L., Schoenmakers, J.G.G.: Eur. J. Biochem. 68, 55 (1976)

Landy, A., Olchowski, E., Ross, W.: Mol. Gen. Genet. 133, 273 (1974)

Marvin, D.A., Hohn, B.: Bacteriol. Rev. 33, 172 (1969)

Miller, J.H.: Experiments in Molecular Genetics. New York: Cold Spring Harbor Laboratory (1972)

Murray, N.E., Murray, K.: Nature (London) 251, 476 (1974)

Rambach, A., Tiollais, P.: Proc. Natl. Acad. Sci. USA 71, 3927 (1974)

Sanger, F., Coulson, A.R.: J. Mol. Biol. 94, 441 (1975)

Thomas, M., Cameron, J.R., Davis, R.W.: Proc. Natl. Acad. Sci. USA 71, 4579 (1974)

Is Integration Essential for Mu Development?

P. van de Putte, M. Giphart-Gassler, T. Goosen, A. van Meeteren, and C. Wijffelman

The genome of bacteriophage Mu can integrate randomly in the chromosome
of its host *Escherichia coli*, thereby causing mutations (Taylor, 1963).
Random integration seems a more widespread phenomenon, since it was
found that other DNA elements also show a more or less random integration: insertion sequences (IS) and transposons (Tn).

However, the frequency of transposition or random integration of these
elements is very low. After Mu infection, when Mu integration functions
are expressed, the random integration has become a very efficient process. The high efficiency of this process will facilitate the elucida-
tion of the mechanism of random integration and of illegitimate recombination events in general.

Comparison Between the Prophage Maps of Mu and λ

Except for its property to integrate very efficiently at random places
of the chromosome, Mu can be considered a normal temperate bacteriophage. In fact Mu (mol wt 24 × 10^6) is very similar to λ (mol wt
30 × 10^6). A comparison between the genetic maps of the two phages will
clarify the differences between Mu and λ and what difference is relevant to the question of random integration.

In Figure 1 the prophage maps of Mu and λ are shown with the genes
having similar functions drawn in corresponding positions. The maps
to the right of the immunity genes are strikingly similar up to the

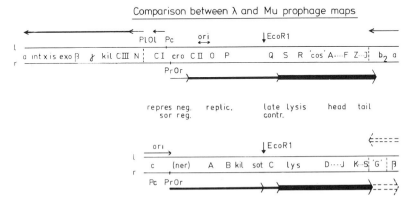

Fig. 1. Comparison between λ and Mu prophage maps. The maps are not drawn to scale,
but with the genes having similar functions drawn in corresponding positions.
Arrows indicate the direction of transcription and replication. Head and tail genes
of Mu are classified on the basis of in vitro complementation

b2 region of λ and the 'G'β region of Mu. Although the b2 region and the 'G'β region are similar in that neither contains essential genes, they do differ structurally. The 'G' area, comprising approximately 8% of the Mu genome, is able to invert due to the presence of inverted repeats, which border the 'G' area (Hsu and Davidson, 1974) and to a 'G' inversion function located on β (Chow et al., 1977). The function of this interesting area remains unclear, but the structure itself does not seem to be unique for Mu, because identical 'G' areas are also found in P 1 and P 7 (Chow et Bukhari, 1977). Thus the total Mu genome resembles the right part of the λ map. As on the right part of the λ genome, Mu transcription is nearly exclusively on the right strand (Bade, 1972; Wijffelman et al., 1974). The part of the λ genome to the left of the immunity gene cI, which is transcribed on the left strand, is lacking in Mu. However, not all the functions located on the left part of the λ genome are absent in Mu.

Let us consider these λ functions, proceeding from the cI gene to the left. First the N gene is met; such a function is not present in Mu (Wijffelman et al., 1974). Similar λ functions, such as cIII, kil, and γ, have been detected in Mu, but are located between genes B and C (Westmaas et al., 1976; unpublished results).

The red genes are absent in Mu; in contrast to λ, no vegetative re-combination is found to occur in a *recA* host (Wijffelman et al., 1972; Couturier and Van Vliet, 1974). Int-like and probably also xis-like functions are present but fundamentally different from λ (see below).

The Structural Differences Between Mu and λ

The Ends of the Mu Genome

Besides the presence of a 'G' area, whose part in integration is not known, there is a very important structural difference between Mu and λ, which certainly is relevant to the integration problem. In contrast to λ DNA, the Mu DNA is not circularly permuted, and the prophage and vegetative maps are the same as shown genetically (Wijffelman et al., 1973; Couturier and Van Vliet, 1974) and by electron-microscopic studies (Hsu and Davidson, 1972). This means that the attachment site for Mu integration is formed by the ends of the linear molecule. Mu does not circularize after infection in the way λ does, because it lacks cohesive ends; on the contrary the ends of the Mu DNA consist of host DNA of varying lengths and composition. For the β-end of the Mu genome this was demonstrated by hetero duplex analysis (Hsu and Davidson, 1972; Daniell et al., 1973). For the immunity end this was shown by restriction enzyme analysis (Allet and Bukhari, 1975) and hybridization (Bukhari et al., 1976). These bacterial ends are lost at the integration process (Hsu and Davidson, 1974).

The Sot Function

Sometimes circularization of a linear DNA molecule, which lacks cohesive ends, can be accomplished by a protein, which keeps the free ends of the DNA molecule together, e.g., as in the *Bacillus subtilis* phage ∅ 29 (Ortin et al., 1971) and in adenovirus (Robinson and Bellett, 1974). The free ends of bacteriophage Mu, however, consist of hetero-geneous DNA, which has to be removed during the integration process. This is probably the reason why such a DNA-protein complex is never found after Mu infection or induction. Nevertheless we have evidence that there is a protein involved in the protection or circularization of Mu. This was derived from transfection studies with Mu DNA. Compared to λ DNA the transfection frequency of Mu DNA is extremely low, but it is strongly stimulated when early Mu proteins are present in the cells before transfection (Table 1). The *A* and *B* gene products,

Table 1. Transfection with Mu DNA

Strain	Relevant characteristics	Transfection frequency[a]
JC 7620	*recBC, sbcB*	$<10^{-7}$
PP 244	*recBC, sbcB, trp*::(Mu *cts*62ΔD-β)	$3,5 \times 10^{-5}$
PP 245	*recBC, sbcB, trp*::(Mu *cts*62A*am*1093)	$2,0 \times 10^{-5}$
PP 246	*recBC, sbcB, trp*::(Mu *cts*62B*am*1066)	$3,5 \times 10^{-5}$

Cells were induced at 43° for 20 min, treated with Ca^{2+}, and Mu DNA (0.1 μg/2×10^8 cells) was taken up at 40° for 10 min. Infective centers were measured on a wild-type *E. coli* K12 strain.

[a]The transfection frequency is given as the number of infective centers per bacterium, present prior to the Ca^{2+} treatment.

being the two known essential early genes, are not responsible for the observed stimulation of transfection. With λ-Mu transducing phages, containing different portions of the early region of Mu, the Sot function (stimulation of transfection) was mapped between genes *B* and *C*, close to the *kil* gene (Van de Putte et al., 1977). Sot⁻ motants can be isolated by selection for polar Kil⁻ mutants. These mutants have an insertion of about 800 base pairs, probably IS1, in the early region of Mu. When polar Kil⁻ mutants were isolated, starting with a complete thermoinducible prophage, the insertions were always found in or before the *B* gene, because besides the *kil* gene, replication must also be blocked to obtain temperature-resistant survivors. The mutants have the same properties as the so-called X mutants (Bukhari, 1975). However, starting with a defective Mu prophage, in which no replication occurs due to a deletion of the β-end, polar Kil⁻ Sot⁻ mutants are found with insertions beyond the *B* gene.

Figure 2 shows a restriction enzyme analysis of a Mu*cts*62, a Mu*cts*62Kil⁻Sot⁻, and a Mu*cts*62X phage using the restriction enzyme EcoRI. Both the Kil⁻Sot⁻ mutant and the X mutant have an insertion in the first EcoR1 fragment of Mu (C fragment). The insertion in the X mutant is in a different HpaI fragment than in the Kil⁻Sot⁻ mutant (results not shown), mapping the latter insertion between 4000 and 5100 base pairs from the immunity end. The Sot function is semi-essential because phages carrying the insertion form pinpoint plaques on Rec⁺ and RecA hosts, but normal plaques on strains containing a polarity suppressor (suA). Because the function is essential for Mu transfection it is possible that the Sot protein is normally injected in the cell in the case of Mu infection. However, this remains to be shown.

Functional Difference Between Mu and λ

Peculiarly, one function is present in λ that is absent in Mu, namely, a nonessential integration function (int). Despite an extensive search for such mutants, only essential genes of Mu are found to be involved

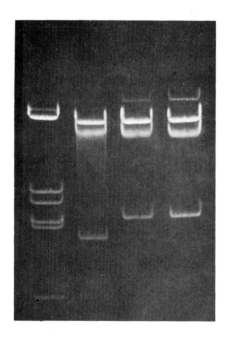

Fig. 2. Agarose gel electrophoresis of EcoRI-generated DNA fragments of Mu DNA. From left to right: λplac5, Muc2000, Mucts62 Kil⁻Sot⁻, and Mucts62X. Both the MuKil⁻Sot⁻ DNA and the MuX DNA have an insertion of about 800 base pairs in the smallest fragment (C fragment)

in integration: the genes *A* and *B* (Van de Putte and Gruijthuijsen, 1972; Toussaint and Faelen, 1973). These genes, however, are also involved in replication (Wijffelman et al., 1974). It is not because of their inability to integrate that *A* and *B* are blocked in replication, because no replication occurs when an *A* or *B* amber prophage is induced in a nonpermissive host (Wijffelman and Lotterman, 1977).

Because of the importance of genes *A* and *B*, we have tried to identify their products in the same way as was done successfully for λ proteins (Hendrix, 1971), namely, by infecting ultraviolet (UV)-irradiated cells with Mu (Giphart and Van de Putte, 1977). In contrast to λ, the amount of Mu proteins made in UV-irradiated cells is extremely low. Moreover only two early proteins are synthesized in sufficient amounts to make their characterization possible on polyacrylamide gels. One has been identified as the *B* gene product (mol wt 35 K), while the other (mol wt 23 K) is lacking in a mutant that is Kil⁻Sot⁻. We have been unable to detect the *A* gene product in this system.

Replication of Bacteriophage Mu

In the previous section we saw that the integration functions coincide with the replication functions. Thus integration might be an essential part of Mu replication. What is known about the replication of Mu? In the first place replication intermediates consisting exclusively of Mu DNA have never been found. On the other hand covalently closed circular DNA is found that consists of Mu and host DNA with a length varying from 1 to 7 times Mu (Schröder et al., 1974; Waggoner et al., 1974). The appearance of these circles is much later than the start of the Mu DNA synthesis, which occurs at about 8 min after induction and infection (Wijffelman and Lotterman, 1977).

In the second place, it has been shown that during the replication cycle, reintegration of Mu DNA takes place with a rather high frequency (Schröder and Van de Putte, 1974; Razzaki and Bukhari, 1975).

More recently we determined the direction of replication on the Mu genome (Wijffelman and Van de Putte, 1977) by hybridizing Okazaki fragments, isolated at 35 min after induction of a Mu-cts prophage, with the separated strands of Mu and of a λ-transducing phage, which contains the early region of Mu. As shown in Table 2a, about 90% of

Table 2. Asymmetric hybridization of pulse-labeled Mu DNA

a: DNA labeled at 35 min after induction

DNA strand on filter	% Strand-specific 'long DNA'	Mu DNA Okazaki fragments
r-Mu	48.5	15.8
l-Mu	51.5	84.2
r-λ Mu	50.0	10.3
l-λ Mu	50.0	89.7

b: DNA labeled at early times after induction

Time	DNA strand on filter	% Strand-specific 'long DNA'	Mu DNA Okazaki fragments
8'	r-Mu	45.3	17.6
	l-Mu	54.7	82.4
10'	r-Mu	47.2	17.7
	l-Mu	52.8	82.3
12'	r-Mu	49.6	16.9
	l-Mu	50.4	83.1

A culture of a Mu-cts lysogen was induced at 43°C. At different times after induction, the culture was labeled for 30 s with [3H]thymidine, lysed, and centrifuged in an alkaline sucrose gradient on a cushion of saturated CsCl. Fractions were assayed for radioactivity, and the 'long DNA' on the CsCl cushion and the Okazaki fragments (±8-12 S) were pooled separately; they were purified by banding in CsCl, after which the DNA-containing fractions were pooled, dialyzed, sonicated, and hybridized to separated DNA strands bound to nitrocellulose filters. The percentage of strand-specific Mu DNA in one fraction is given by the ratio: radioactivity bound to one strand/radioactivity bound to both strands x 100

the Okazaki fragments hybridize with the l-strand of Mu and of the λ-Mu transducing phage and only about 10% with the r-strands, while such an asymmetry in hybridization is absent with the high-molecular-weight DNA. The same asymmetry was found with fragments isolated during the first round of replication (Table 2b).

This means that even in the first round the replication of Mu is unidirectional for at least 90%. However, considering comparable data of P2 (Kainuma-Kuroda and Okazaki, 1975), it is likely that the replication of Mu is completely unidirectional.

The direction of replication of Mu is from the immunity end toward β as follows from the polarity of the DNA strands (Fig. 3), which was previously determined by transciption studies (Wijffelman and Van de

Genetic map

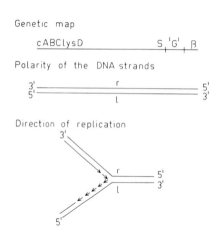

Fig. 3. Genetic map showing direction of replication of Mu

Putte, 1974). Assuming the absence of a normal excision after induction of a Mu prophage, it is obvious that Mu DNA synthesis is initiated at or very near the immunity end. The start of replication is probably at the immunity end, because there is no decrease in asymmetry when Okazaki fragments are hybridized with the early region of Mu, present in the λ Mu transducing phages (Table 2a). However, the β-end of Mu is essential for replication: A defective prophage, lacking the β-region and part of the 'G'-area, does not show any replication at all. So the exciting possibility arises that the origin of replication is formed by the ends of Mu and might be the same as the attachement site for integration.

When replication proceeds from the immunity- to the β-end, the formation of the Mu coli circles could be explained by replication that moves on beyond the β-end into the bacterial chromosome. This possibility was studied in a Mu lysogen, in which Mu was integrated in a λ prophage between the λ genes A and D. After induction of Mu the newly synthesized DNA was labeled and analyzed by hybridization with Mu and λ DNA (Fig. 4). No increase in λ DNA synthesis could be detected. For this reason it is unlikely that Mu replication proceeds to an appreciable extent into the regions adjacent to the Mu prophage. Therefore the formation of the Mu coli DNA circles cannot be explained in this way. An alternative explanation could be that their formation is due to recombination and not to replication. A model explaining the replication data obtained thus far is shown in Figure 5: 1) Starting with a prophage, the origin of replication is formed by joining the two ends of Mu by means of proteins. 2) This complex is activated by a single-stranded break; right afterwards, replication starts at the immunity end and proceeds to the right, where it is terminated at or just outside the β-end. In this way one copy remains in the chromosome and the other is "peeled off." 3) The ends of the "peeled-off" copy are joined again by proteins, but replication cannot start again unless this Mu copy is reintegrated. This simple model not only explains why Mu needs many integration sites, but also why Mu integration and replication functions are coupled: Integration is an essential part of the replication machinery.

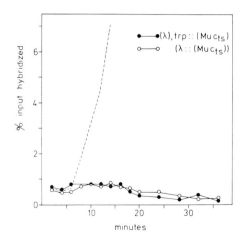

Fig. 4. Replication of Mu and λDNA after induction of a Mu*cts* prophage inside a λ prophage. After growth at 30°, the strains were induced and kept at 42°. At different times after induction the DNA was pulse-labeled with [³H]thymidine for 1 min. Then the cells were lysed and the DNA extracted with phenol, sonicated, and denatured. The DNA was hybridized with filters loaded with λ*b2* DNA. The data are given as the ratio cpm hypridized/TCA-precipitable input x 100%. The dotted line reflects the Mu DNA synthesis; in this case samples were hybridized with Mu DNA loaded on the filter

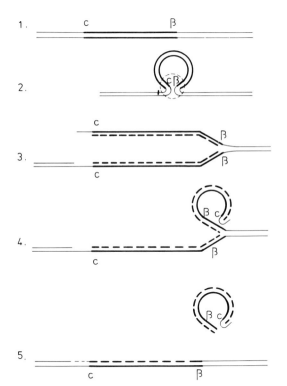

Fig. 5. Model of the first events taking place after Mu induction. The heavy lines represent Mu DNA. The ends of Mu are indicated by c (immunity end) and β (the β end)

References

Allet, B., Bukhari, A.: J. Mol. Biol. 92, 529-540 (1975)
Bade, E.: J. Virol. 10, 1205-1207 (1972)
Bukhari, A.: J. Mol. Biol. 96, 87-99 (1975)
Bukhari, A., Froshauer, S., Botchan, M.: Nature (London) 264, 580-583 (1976)

Chow, L., Bukhari, A.: In: •DNA Insertion Elements, Plasmids and Episomes. Bukhari, A., Shapiro, T., Adhya, S. (eds.). New York: Cold Spring Harbor Laboratory, 1977, pp- OO-OO

Chow, L., Kahman, R., Kamp, D.: In: DNA Insertion Elements, Plasmids and Episomes. Bukhari, A., Shapiro, T., Adhya, S. (eds.). New York: Cold Spring Harbor Laboratory, 1977, pp. OO-OO

Couturier, M., Van Vliet, F.: Virology 60, 1-8 (1974)

Daniell, E., Abelson, J., Kim, J., Davidson, N.: Virology 51, 237-239 (1973)

Giphart, M., Van de Putte, P.: T. Mol. Biol. in press (1977)

Hendrix, R.: In: The Bacteriophage Lambda Hershey, A.D. (ed.). New York: Cold Spring Harbor Laboratory, 1971, pp. 355-370

Hsu, M., Davidson, N.: Proc. Natl. Acad. Sci. USA 69, 2823-2827 (1972)

Hsu, M., Davidson, N.: Virology 58, 229-239 (1974)

Kainuma-Kuroda, R., Okazaki, R.: J. Mol. Biol. 94, 213-228 (1975)

Ortin, J., Viñuela, E., Salas, M.: Nature (London), New Biol. 234, 275-277 (1971)

Razzaki, T., Bukhari, A.: J. Bacteriol. 122, 437-442 (1975)

Robinson, A., Bellett, A.: Cold Spring Harbon Symp. Quant. Biol. 39, 523-531 (1974)

Schröder, W., Bade, E., Delius, H.: Virology 60, 534-542 (1974)

Schröder, W., Van de Putte, P.: Mol. Gen. Genet. 130, 99-104 (1974)

Taylor, A.: Proc. Natl. Acad. Sci. USA 50, 1043-1051 (1963)

Toussaint, A., Faelen, M.: Nature (London) New Biol. 242, 1-4 (1973)

Van de Putte, P., Gruijthuijsen, M.: Mol. Gen. Genet. 118, 173-183 (1972)

Van de Putte, P., Westmaas, G., Gassler, M., Wijffelman, C.: In: DNA Insertion Elements, Plasmids and Episomes. (Bukhari, A. Shapiro, T., Adhya, S. (eds.). New York: Cold Spring Harbor Laboratory, 1977, pp. OOO-OOO

Waggoner, B., Gonzalez, N., Taylor, A.: Proc. Natl. Acad. Sci. USA 71, 1255-1259 (1974)

Westmaas, G., Van der Maas, W., Van de Putte, P.: Mol. Gen. Genet. 145, 81-87 (1976)

Wijffelman, C., Gassler, M., Stevens, W., Van de Putte, P.: Mol. Gen. Genet. 131, 85-96 (1974)

Wijffelman, C., Lotterman, B.: Mol. Gen. Genet. 151, 169-174 (1977)

Wijffelman, C., Van de Putte, P.: Mol. Gen. Genet. 135, 327-337 (1974)

Wijffelman, C., Van de Putte, P.: In: DNA Insertion Elements, Plasmids and Episomes. Bukhari, A., Shapiro, T., Adhya, S. (eds.). New York: Cold Spring Harbor Laboratory, 1977, pp. OOO-OOO

Wijffelman, C., Westmaas, G., Van de Putte, P.: Mol. Gen. Genet. 116, 40-46 (1972)

Wijffelman, C., Westmaas, G., Van de Putte, P.: Virology 54, 125-134 (1973)

The Role of IS-Elements in *E. coli*

H. Saedler and D. Ghosal

The chromosome of *Escherichia coli* is a circular double-stranded DNA molecule with a length of about 1 mm, which is folded in a unique way. This DNA molecule has a coding capacity of some 3000-4000 average-sized polypeptide chains. The genes coding for these proteins are thought to be arranged in the chromosome in a sequence characteristic for a given bacterial species.

However, in recent years genetic elements have been detected that can cause considerable disturbance in the arrangement of genes. These special DNA sequences are called IS (insertion sequence) elements. They were shown to be natural components of the *E. coli* chromosome. This article is a review of some of these properties and contains a brief outline of the role of IS elements in evolution.

Detection of IS Elements

The presence of IS elements in the chromosome of *E. coli* was originally detected through their property of transposing from one position in the chromosome to another, thus frequently inducing mutations. Further analysis of these mutations led to our present understanding of IS elements (for review, see Ref. 1).

Strong Polarity

The integration of an IS element into a gene of an operon not only inactivates that particular gene but also abolishes the expression of the more promoter distal genes. IS-induced mutations are therefore said to be strongly polar (2).

Therefore they seem to carry signals interfering with gene expression. In vitro studies have shown that at least some IS elements carry a termination signal for RNA polymerase (3). In the presence of the bacterial rho protein transcription is terminated at these sites (thus genes further downstream are not transcribed and hence not translated).

Physical Evidence for Insertion of DNA

Many different lines of experiments showed that the strongly polar mutations are due to the integration of DNA into the continuity of a gene.

Inspection of DNA Heteroduplex Molecules in the Electron Microscope

This technique is illustrated in Figure 1. One DNA strand of a restriction fragment carrying a strongly polar mutation in galT is hybridized

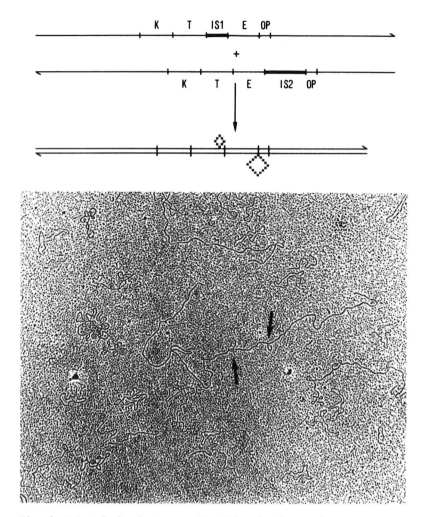

Fig. 1. Heteroduplex between galT::IS1 and galOP::IS2 DNA strands

with a complementary DNA strand of a similar fragment but carrying a
strongly polar mutation in the control region of the galactose operon.
A small and a large single-stranded DNA loop are seen at the position
of the mutations. The smaller loop is 800 base pairs long, while the
larger is about 1400 base pairs long.

Analysis of many such strongly polar mutations in all possible combinations led to two important findings (4 - 6): 1) A few discrete size
classes of inserted DNA are observed, and 2) these fall into various
homology groups. For example: All the small insertions have the same
homology, if hybridized with each other. Since these DNA pieces are
never observed to be circularly permutated in their DNA sequence,
they apparently are transposed as units. Therefore they can be considered as elements.

Table 1 contains a list of all known transposable IS elements in *E.
coli*.

Table 1. Transposable elements in bacteria

Name	Size	Terminal repetitions	Associated drug resistance	Effect on gene expression
IS1	730	–	–	polar (I/II)
IS2	1 350	–	–	polar (I), constitutive (II)
IS3	1 400	–	–	polar (I)
IS4	1 400	–	–	polar (I/II)
IS5	1 400	–	–	?
IS6	115	–	–	constitutive (I)
IS7	60	–	–	constitutive (I)
Tn1	4 800	140, IR	Amp	polar (I)
Tn2	4 800	"	Amp	polar (I)
Tn3	4 600	"	Amp	–
Tn4	20 500	"	Amp. Sul, Str	–
Tn5	5 200	1 400, IR	Kan	polar (I/II)
Tn6	4 100	–	Kan	–
Tn7	13 500	–	Tmp, Str	–
Tn9	2 500	800, DR, IS1	Cam	polar (I/II)
Tn10	8 500	1 400, IR, IS3	Tet	polar (I/II)

Inspection of DNA heteroduplex molecules in the electron microscope, however, is limited in resolution to certain sizes of DNA insertions. If the size of an IS element is an order of magnitude smaller than, e.g., IS1 through IS5 (see Table 1), the heteroduplex technique is not sensitive enough to reveal the presence of such smaller insertions. To analyze these mutations, another technique seems to be more adequate.

Comparative Analysis of DNA Fragments Generated by Various Restriction Enzymes in Gel Electrophoresis

If suitable restriction fragments are available from a gene that carrys and one that does not carry an integrated IS element, electro-phoresis of these fragments in agarose gels will show the difference in molecular weight due to the integrated IS element.

We subjected to gel electrophoresis DNA fragments of two mutants, which, based on genetic criteria, we had reason to believe were due to the integration of DNA (Ghosal and Saedler, manuscript in preparation).

Figure 2 shows the pattern of a HindII, HindIII double digest of various plasmid DNAs.

Slot 3 shows the pattern of the parental plasmid pDG1, which is gal[+]. Slot 4 gives the pattern of pDG12, in which an IS2 is integrated in the control region of the gal operon (see Fig. 1); thus, a cell carrying this plasmid has a gal[−] phenotype. Note the appearance of 2 new bands (e and f) and the shift in molecular weight of one band (from a to b), due to the integration of IS2.

Slots 1 and 5 show the pattern of two independent gal[+] revertants obtained from plasmid pDG12. Note the increase in molecular weight of only band e in both mutants. This can only be interpreted if an insertion has occurred in band e, thus not affecting other bands. Using the markers (slot 2) as references, the increase in molecular weight can be calculated. Mutation 1 (slot 5) is due to the integration of about

44

a 115 base pair-long piece of DNA, while the other mutation (slot 1) is about a 60 base pair insertion. The former is given the name IS6 and the latter is termed IS7.

Both insertions confer a gal⁺ phenotype to the cells carrying the plasmid. Since they seem to have integrated into IS2 they either destroy the polar signal on IS2 or, what is more likely, they carry their own turn on signal.

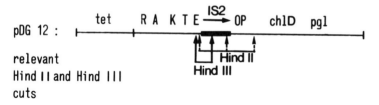

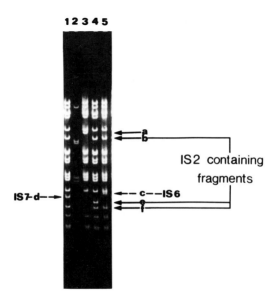

Fig. 2. Identification of the mini-insertions IS6 and IS7

Turn-off and Turn-on of Gene Activity by IS2

As seen in Figure 2, IS2 turns off gene activity in orientation I. If, however, IS2 is integrated in orientation II, the adjacent genes are turned on very efficiently, allowing 2 - 3 times more transcription of the operon than through the normal gal promoter (7).

Apparently IS2 carries a promoter, which, if the element is integrated in orientation II, allows transcription of the operon, while in orientation I the system is shut off. Recently Rak (8) showed that mRNA synthesis is initiated within IS2.

This IS element, therefore, can be considered as a prefabricated element suitable in evolution since it is transposable and carries a promoter and other signals necessary for the turning on and turning off of other gene activities.

IS6 and IS7 seem to be similar in this respect. At least in one orientation they turn on genes. Their effect on gene expression in the other orientation is not known.

Unlike IS2, the much smaller IS6 and IS7 are highly unstable, leading us to another property of IS elements.

Accurate Excision of IS Elements

Most strongly polar mutations revert spontaneously. The frequency of reversion cannot be increased by mutagens (9). This seems to indicate that the integrated IS element is excised accurately, restoring the wild-type genotype. Since excision occurs equally well in bacteria deficient in normal recombination, it implies that the normal recombinational pathway is not involved in excision of IS elements (10,11).

No positive evidence is available about the enzyme involved in integration and excision of IS elements. Excision of IS elements, however, is not always precise.

IS-induced Deletions

Another type of recombinational event has been reported (11), involving the termini of an integrated IS element and leading to the loss of genetic material adjacent to the IS element.

The frequency of IS1-induced deletion formation is independent of recA, as is accurate excision of IS1. Conversely to accurate excision of IS1, deletion formation mediated by IS1 shows a strong dependence on the growth temperature of the cells. Since IS1 is seldom lost in this process, the instability is conserved and further rounds of deletion formation can be triggered (12). Recently we isolated a mutant deficient in this process (13).

Apparently IS1 can serve as a generator for gene rearrangements by fusing genes that previously were separated. This process can lead to new combinations of genes and expression signals and thus might be attractive in evolution (11).

The fusion of genes is especially appealing if no genetic information is lost during this process, as is the case if a duplication has occurred prior to fusion.

Duplication Mediated by IS2

Duplications can arise by a single recombinational event if the chromosome is replicating. In one daughter molecule, some genetic material is duplicated while it is lost from the other chromosome. Deletions and duplications therefore would be correlated in such an event.

There is genetic evidence that IS2 seems to mediate duplications (14).

The deletion fusions of genes and expression signals can potentially yield new regulator circuits. Transpositions of genes from one site to another can also lead to rearrangements of potential value.

Transposition of Genes Mediated by IS Elements

From what has been said so far, it is clear that IS elements are involved in recombinational events causing deletions and duplications. The termini of the integrated IS elements seem to be the sites at which the recombinational events occur (11).

The IS-elements themselves are transposable. In addition, they seem to be able to function as vectors for other genetic material to move from one site in the cromosome to another. If a constant length of DNA coding for a particular function is always transposed as a unit, physical analysis of these so-called transposons has shown that they are flanked by special duplicated DNA sequences either in direct or inverted orientations (15). In at least two such cases, these flanking sequences are known IS elements (16, 17). The known transposons are included in Table 1.

Is elements therefore seem to stimulate gene rearrangements by inducing deletions in their vicinity causing duplication of genetic material and finally to transpose other genes to other sites of the same or other DNA molecules.

These processes are thought to be important in evolution. The question arises concerning the origin of IS elements. Are they invaders like bacteriophages or are they normal constituents of chromosomes.

IS Elements Are Natural Components of the *E. Coli* Chromosome

Some years ago we showed by DNA-DNA hybridization that at least IS1 and IS2 are natural components of the *E. coli* chromosome, where they occur in multiple copies. We found about eight copies of IS1 and about five copies of IS2 per chromosome (18).

The accurate positions of the various copies in the chromosome and their orientations are unknown, but in principle they can occur either in direct or inverted orientations. The latter can be seen in the electron microscope if chromosomal DNA is denatured and rapidly renatured at low DNA concentrations. If two copies of identical DNA sequences occur in the chromosome in inverted orientations they should hybridize with each other under the foregoing conditions, forming a double-stranded DNA stem terminating in a single-stranded DNA loop. The size of the stem is indicative of the IS element involved, and the length of the loop is a measure of the distance between the duplicated genetic material. Figure 3 shows three different examples of this kind. The stem in the right-hand picture is of the size of IS1, and the loop is about 22 kilo base pairs long. The stem in the middle picture is equal

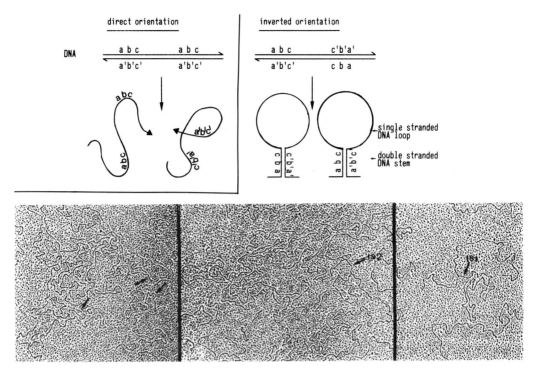

Fig. 3. Duplication of genetic material in the chromosome of *E. coli* K 12

to the size of IS2, IS3, IS4, or IS5; the loop is about 120 kilo base pairs long. In the molecule shown in the left-hand picture, multiple interactions are seen with lengths varying from 100 to 300 base pairs. The smaller ones might correspond to IS6.

However, none of these stems has been proved to be homologous to one of the known IS elements. A more detailed analysis using this technique has been done by Chow (19) and Deonier and Hadley (20), showing that the stems and the loops fall into discrete size classes. This could indicate that the multiple copies of IS elements are not distributed randomly in the *E. coli* chromosome.

Conclusions

IS elements are natural components of the *E. coli* chromosome. Besides stimulating a number of illegitimate recombinational events like duplication, deletion, and transposition, which are considered to be important in evolution, they also carry signals necessary for gene expression. Similar elements are also known in higher organisms (Nevers and Saedler, manuscript in preparation).

References

1. Starlinger, P., Saedler, H.: Curr. Top. Microbiol. Immunol. 75, 111 (1976)
2. Starlinger, P., Saedler, H.: Biochemie 54, 177 (1972)
3. De Crombrugghe, B., Adhya, S., Gottesman, M., Pastan, H.A.: Nature (London), New Biol. 241, 260 (1973)
4. Hirsch, H.J., Starlinger, P., Brachet, P.: Mol. Gen. Genet. 119, 191 (1972)
5. Fiandt, M., Szybalski, W., Malamy, M.H.: Mol. Gen. Genet. 119, 223 (1972)
6. Saedler, H., Kubai, D., Nomura, M., Jaskunas, S.R.: Mol. Gen. Genet. 141, 85 (1975)
7. Saedler, H., Reif, H.J., Hu, S., Davidson, N.: Mol. Gen. Genet. 132, 265 (1974)
8. Rak, B.: Mol. Gen. Genet. 149, 135 (1976)
9. Saedler, H., Starlinger, P.: Mol. Gen. Genet. 100, 178 (1967)
10. Jordan, E., Saedler, H., Starlinger, P.: Mol. Gen. Genet. 100, 296 (1967)
11. Reif, H.J., Saedler, H.: Mol. Gen. Genet. 137, 17 (1975)
12. Reif, H.J., Saedler, H.: In: DNA Insertion Elements, Plasmids and Episomes. Bukhari, A.I., Shapiro, J., Adhya, S. (eds.) New York: Cold Spring Harbor Laboratory, 1977, in press
13. Nevers, P., Reif, H.J., Saedler, H.: In: DNA Insertion Elements, Plasmids and Episomes. Bukhari, A.I., Shapiro, J., Adhya, S. (eds.) New York: Cold Spring Harbor Laboratory, 1977, in press
14. Ahmed, A.: Mol. Gen. Genet. 136, 243 (1975)
15. Cohen, S.N.: Nature (London) 263, 731 (1976)
16. Ptashne, K., Cohen, S.N.: J. Bacteriol. 122, 776 (1975)
17. MacHattie, L., Jackowski: In: DNA Insertion Elements, Plasmids and Episomes. Bukhari, A.J., Shapiro, J., Adhya, S. (eds.). New York: Cold Spring Harbor Laboratory, 1977, in press
18. Saedler, H., Heiß, B.: Mol. Gen. Genet. 122, 267 (1973)
19. Chow, L.T.: In: DNA Insertion Elements, Plasmids and Episomes. Bukhari, A.I., Shapiro, J., Adhya, S. (eds.). New York: Cold Spring Harbor Laboratory, 1977, in press
20. Deonier, R.C., Hadley, R.G.: Nature (London) 264, 191 (1976)

DNA Recombination in Cells Infected with Simian Virus 40

D. Nathans and M. W. Gutai

Introduction

The genome of Simian Virus 40 (SV40) is a covalently closed circular
DNA duplex with about 5200 nucleotide pairs. Based on physiological
studies and mapping of temperature-sensitive and deletion mutants,
three genes have so far been identified (see Fig. 1). The A gene maps

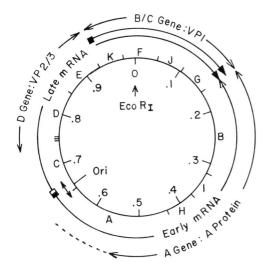

Fig. 1. Map of the SV40 genome. The
EcoR$_I$ cleavage site is the zero point
and map units are as indicated clock-
wise from this site. *Ori;* origin of
DNA replication

in the early region defined by mRNA mapping, and codes for a protein
estimated by gel electrophoresis to be about 90,000 daltons that has
"tumor antigen" determinants. The B/C gene maps between 0.94 and 0.17
map units in the late region of the genome and codes for the major
structural protein VP1. The D gene maps roughly between map coordinates
0.76 and 0.96, also in the late region, and codes in phase for the
overlapping VP2 and VP3 minor virion proteins. (For a recent review
of the SV40 genome, see Kelly and Nathans, 1977.)

During productive infection of monkey cells with SV40 there is extensive replication of viral DNA. Each molecule replicates bidirectionally
from a unique site (ori) located at 0.67 map units (see Fig. 1); replication terminates 180° from the origin. The terminus is not determined by a nucleotide sequence, but is the point where the two replication forks meet. Therefore the ori sequence is the only cis element
required for SV40 DNA replication.

In addition to replication and development of progeny virus in SV40-
infected permissive cells, extensive recombination occurs between and
within SV40 DNA molecules and between SV40 and cellular DNA. A dia-

grammatic listing of recombination reactions involving SV40 DNA is
shown in Figure 2. We shall concentrate on a few of these reactions,
namely, generation of deletions and duplications, and recombination
of SV40 DNA with cellular DNA resulting in substituted variants or
integration. In particular, we want to focus on the possible specificity of recombination sites within the SV40 genome and in cellular
DNA.

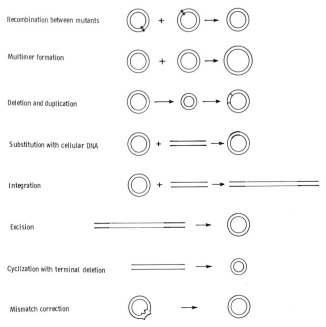

Fig. 2. Diagrammatic listing of recombination events detected in SV40-infected cells

Recombinant Variants of SV40

The results of recombination events involving SV40 DNA is most dramatically seen during serial passage of the virus in permissive cells at
high multiplicity of infection (Uchida et al., 1968; Tai et al., 1972).
Under these conditions, each cell is multiply infected and therefore
wild type virus can supply all trans functions. Any defective recombinant DNA molecule can propagate and be retained during passage,
provided it has kept the ori signal and any nucleotide sequence required for encapsidation. In addition, to be efficiently encapsidated,
the molecule must be between about 70 and 100% of the length of wild
type SV40 DNA.

Figure 3 shows the evolution of SV40 recombinant variants during serial passage, in the form of an electropherogram of viral DNA species
at every fifth passage up to passage 60. As seen in the figure, short
DNA species (i.e., with greater electrophoretic mobility than wild
type DNA) are abundant and sometimes predominate, and during later
passages there are only a few stable size classes of viral genomes. As
discussed more extensively elsewhere (Lee et al., 1975), new species

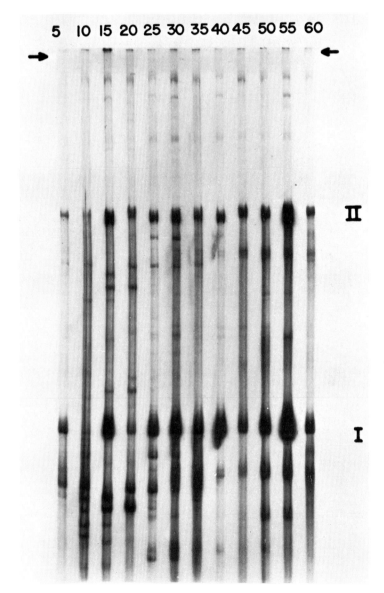

Fig. 3. Evolution of SV40 variants during serial passage revealed by agarose gel electrophoresis of viral DNA. Shown is an autoradiogram of ^{32}P viral DNA extracted from cells infected with undiluted lysates of 5th passage virus *(5)*, 10th passage virus *(10)*, etc. *I:* position of wild type form I SV40 DNA; *II:* position of wild type form II SV40 DNA. Electrophoresis was from top *(arrows)* to bottom

of DNA evolve with selective replication advantage based on multiple ori signals for DNA replication. As discovered by Winocour and his colleagues (Lavi and Winocour, 1972), some of these new species contain cellular DNA sequences covalently linked to SV40 DNA. Based on DNA annealing experiments, illustrated in Figure 4 and in Frenkel et al. (1974) and Davoli et al. (1977), both highly reiterated cell sequences and less reiterated sequences are detectable.

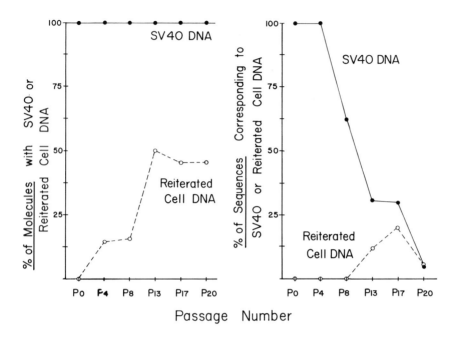

Passage Number

Fig. 4. The occurrence of SV40 DNA sequences and cellular DNA sequences in unfractioned viral DNA extracted from cells infected with high passage lysates. ^{32}P labeled high passage viral DNA was nicked and denatured, and annealed with excess SV40 or cellular DNA, the latter under conditions that lead to annealing of highly reiterated sequences only. *Percent of molecules* containing any SV40 or highly reiterated cellular DNA sequences was determined by hydroxyapatite adsorption. *Percent of sequences* representing SV40 or highly reiterated cellular DNA sequences was determined by resistance to S1 nuclease. For details see Brockman et al. (1973)

To determine the precise structure of SV40 recombinant genomes, several variants have been cloned by plaque formation in the presence of a ts or wild type helper virus and analyzed by restriction enzyme mapping, electron microscopic heteroduplex mapping, and where appropriate, by DNA annealing experiments, as detailed elsewhere (Brokman and Nathans, 1974; Mertz et al., 1974; Brockman et al., 1975). In general four types of variants have been found: (1) viable deletion mutants; (2) complementing variants that retain one or two SV40 genes and no cellular DNA; (3) grossly rearranged viral genomes with no intact genes and no cellular DNA; and (4) substituted variants that retain only a small segment of SV40 DNA, including the ori sequence, joined to cellular DNA. Examples of each of these classes of variants are shown in Figures 5 and 6. Shown in Figure 5 is an example of a viable deletion mutant [pm(d1)810] (Mertz et al., 1974) and two complementing variant genomes each with a deletion and a duplication (the latter including the ori sequence) and therefore two recombinant joints (Brockman et al., 1975). Shown in Figure 6 (DAR variant) is a grossly rearranged genome containing an inversion and tandem repetition of SV40 DNA segments. Also shown in Figure 6 are two examples of substituted variant genomes, each consisting of a tandem repeat of a segment of DNA that has an SV40 ori sequence joined to cellular DNA (Lee et al., 1975).

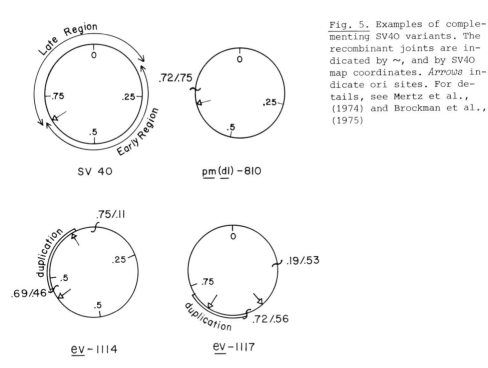

Fig. 5. Examples of complementing SV40 variants. The recombinant joints are indicated by ~, and by SV40 map coordinates. *Arrows* indicate ori sites. For details, see Mertz et al., (1974) and Brockman et al., (1975)

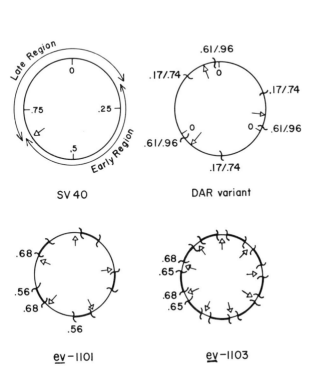

Fig. 6. Examples of variants with grossly altered structure, some containing substitutions of cellular DNA. Symbols as in Figure 5. *Heavy lines* in ev-1101 and ev-1103 indicate cellular DNA. For details, see Khoury et al. (1974), and Lee et al. (1975)

Specificity of SV40 Recombination Sites

Are the recombinant joints of evolutionary variants located at specific sites in the SV40 genome? In Figure 7a and Table 1 are shown the
SV40 map positions of recombinant joints in the variant genomes shown
in Figures 5 and 6 and in other recombinants described in the literature.
(Also illustrated in Figure 7 are positions of recombinant joints in
integrated genomes and in adeno-SV40 hybrid viruses, to be discussed
below.) As seen in the figure, there are many different sites in SV40
DNA where recombination has occurred. Some of these (particularly the
joints related to duplications and the substituted variant joints between SV40 and cellular DNA) are clustered around the origin of replication at 0.67 map units. This is probably due to the selection of variants
with multiple origins, rather than recombination site specificity,
since even within this group there are differences in these sites.
Moreover, some of the substituted variants come from the same evolutionary series, and therefore one may be derived from another by sequential recombination events. Our general conclusion is that any of
a number of SV40 sites, where site is defined in terms of SV40 map
coordinates, (and perhaps any site) can participate in intra- or
intermolecular recombination.

   a. Evolutionary Variants   b. Adeno-SV40 Hybrids   c. Integrated SV40 DNA

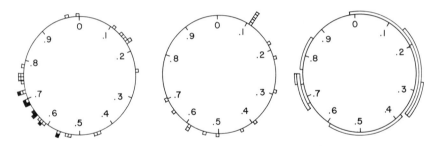

Fig. 7a - c. SV40 map positions of joints found in recombinants. (a) Evolutionary
variants. Each *open square* indicates the position of a SV40-SV40 DNA joint. Each
closed square represents the position of a SV40-cell DNA joint. (Data from Khoury
et al., 1974; Mertz et al., 1974; Brockman et al., 1975; Lee et al., 1975; Davoli
et al., 1977; Gutai and Nathans, unpublished observations.) (b) Adeno-SV40 hybrids.
Each square indicates the position of a SV40-Ad2 DNA joint. (Data from Kelly and
Lewis, 1973; Morrow et al., 1973; Lebowitz et al., 1974; Kelly et al., 1974; Kelly,
1975.) (c) Integrated SV40 DNA in transformed cells. Each bar indicates the limits
of the position of a SV40-cell DNA joint in a given transformed cell. (Data from
Botchan et al., 1976; Ketner and Kelly, cited in Kelly and Nathans, 1977)

At the nucleotide sequence level, our recent data (Gutai and Nathans,
unpublished observations) indicate that each of six independently
derived joints in a substituted variant genome is made up of a distinc-
tive SV40 (and cellular) DNA sequence; two such joints are shown in
Figure 8. Therefore, no single nucleotide sequence serves as the signal
for SV40 recombination. Further, if sequence-specific recombination is
involved in the generation of variants, there must be several such
systems in the cell.

BASE SEQUENCE AT TWO SV40-CELL DNA JOINTS

SV40   Cell

5' ---CGGAACTGTGACTCATCACT--- 3'

5' --- CTAGGTCAACTGGGTATCT --- 3'

Fig. 8. Nucleotide sequences at two SV40-cellular DNA joints of a substituted variant (Gutai and Nathans, unpublished observations)

Given that there are many sites in SV40 or in cellular DNA where re-combination may occur, are there preferred sites? Hsu and Jelinek (1977) and Shen and Hearst (1977) have suggested on the basis of in-trastrand base pairing of SV40 DNA that inversions of nucleotide se-quences at various positions are related to intramolecular recombina-tion resulting in deletions, duplications and inversions of DNA. How-ever, until joints of variant genomes are analyzed at the sequence level and compared to the inverted sequences inferred by microscopy, their correspondence will remain conjectural. Other evidence suggesting preferred SV40 sites for intramolecular recombination are the findings by Chow et al. (1974) and by Davoli et al. (1977), that identical or nearly identical variants containing deletions, inversions, or sub-stitutions appear in independently generated virus stocks. Again, the actual identity of the recombination sites will need to be established by sequence analysis.

Specificity of Cellular Sequences

In the case of cellular DNA sequences in substituted variants, both highly reiterated cell DNA sequences and less reiterated or nonreiter-ated sequences have been found as indicated earlier (Fig. 4); (Brock-man et al., 1973; Rozenblatt et al., 1973; Frenkel et al., 1974; Davoli et al., 1977; Rosenberg et al., 1977). The two cloned substituted variants analyzed by Lee et al. (1975; those shown in Fig. 6) do not share cellular DNA sequences, as judged by DNA hybridization experi-ments: ev-1101 has no highly repetitive cellular DNA sequences, whereas ev-1103 has both highly repetitive sequences and rarer cellular se-quences. However, Winocour and his colleagues (Oren et al., 1976) have clearly shown by cross-hybridization that some, but not all, in-dependently generated variants share certain cellular DNA sequences, including nonrepetitive sequences. More recently, Rosenberg et al. (1977) reported that cross-hybridizing cellular sequences, corres-ponding by nucleotide sequence analysis to highly repetitive cell DNA, are present in a series of independently derived substituted variants, including variants isolated in different laboratories. We have just confirmed this finding by nucleotide sequence analysis of an additio-nal cloned variant.

These several results indicate that different types of cellular DNA can be joined to SV40 DNA during productive infection, but that cer-tain cellular DNA segments are found linked to SV40 with unexpectedly high frequency, including cell DNA that is not highly reiterated in the cell. Whether the frequent occurrence of particular cellular DNA sequences is due to specificity of recombination, to the physical

state of the cellular DNA involved in recombination, to selective
advantage of such sequences, or to other causes, is not known. However, based on the fact that cellular nucleotide sequences are different at different SV40-cell DNA recombinant joints (as noted above,
Fig. 8), we reiterate the conclusion that sequence-specific recombination would require several different enzyme systems.

Adeno-SV40 Hybrid Genomes

Another well-studied example of SV40 DNA recombination is the generation of adeno-SV40 hybrid viruses in monkey cells co-infected with
SV40 and human adenovirus (Lewis et al., 1974). A summary of SV40 recombination sites found in a series of such hybrids is shown in Figure
7b, based on mapping studies of Kelly and Lewis (1973); Morrow et al.
(1973); Lebowitz et al. (1974); Kelly et al. (1974) and Kelly (1975).
(It should be noted that several hybrids containing a recombinant
joint at 0.11 SV40 map units - see Fig. 7b - were probably derived
from a single parent.) From the multiplicity of SV40 sites found in
the hybrid genomes it is apparent that recombination with adenovirus
DNA can occur at several sites on the SV40 genome.

Integrative Recombination

During cell transformation by SV40, recombination between viral and
cellular DNA results in integration of viral genes into cellular DNA.
At present it is not known wheter integrative recombination leading
to transformation and recombination detected during productive infection differ mechanistically. SV40 integration (recombination) sites
have been mapped by restriction enzyme analysis of transformed cell
DNA at many sites in the SV40 genome (see Fig. 7c), and also at different sites in cellular DNA (Botchan et al., 1976; Ketner and Kelly,
1976 and cited in Kelly and Nathans, 1977). Nothing is yet known,
however, about the nucleotide sequence specificity of integrative recombination.

Discussion

In SV40-infected cells recombination is readily detectable within
SV40 DNA molecules and between SV40 DNA and other DNAs present in the
cell nucleus: cellular DNA, other SV40 DNA molecules, or other viral
DNAs. At the map coordinate level many different sites in the SV40
DNA molecule appear capable of recombining; and different regions in
cellular DNA can also be involved. In the case of evolutionary variants, both highly reiterated cell DNA sequences and less reiterated
sequences have been detected. However, it is clear that certain cell
DNA sequences appear in variant genomes with a frequency higher than
expected. Although the basis for this high frequency is not known, it
could be due to differences in the physical state of certain regions
of cellular chromosomal DNA, allowing more frequent recombination
with viral DNA, or perhaps to selective advantage of particular cell
DNA sequences.

At the nucleotide sequence level, knowledge is more limited, but our recent sequence analysis of six different SV40-cell DNA joints in substituted variants indicates that many different viral and cell sequences can be involved. Therefore, if the observed recombination is nucleotide sequence-specific, a series of site-specific enzymes would be required. Our data suggest another basis for recombination at many sites on SV40 and cellular DNA, namely, the formation of a recombinant joint with limited, perhaps patchy, base sequence homology. Indeed, nucleotide sequence data at several of the SV40-cell DNA joints of one of our substituted variant genomes suggest that patchy sequence homology could underlie formation of the initial recombinant, as illustrated for one such joint in Figure 9. According to this hypothesis, the

HYPOTHETICAL JOINT WITH PATCHY HOMOLOGY

Recombinant Sequence

```
         TGACTCATCTC        AACCGCT  T      CCAG        T  CCAG

5'-ACTG             AGGGGC        GC GAGTT    ATGGTTG TG

3'-TGAC             TCCCCG        CG CTCAA    TACCAAC AC

         CCGCCTCAA        CCCTACC  C      T   A       G  TGAT
                                          C   G
                                          C   T
                                          C   C
                                          C   C
                                          GC
```

SV40 Sequence

Fig. 9. A model for formation of recombinant joints based on patchy sequence homology. The nucleotide sequences shown are those of SV40 *(bottom)* and a recombinant joint present in a substituted variant *(above)*. SV40 sequence from Subramanian et al. (1977); recombinant sequence from Gutai and Nathans, unpublished observations

initial joint molecule is formed by interrupted or discontinuous patches of base pairs (Fig. 9), following which excision, repair, and ligation result in the final recombinant joint sequence, which may be derived from one or both parents. Further analysis of recombinant joints, particularly those derived from parental DNA segments of known sequence should help determine the role of sequence homology in SV40 recombination.

A broader question is whether recombination seen in SV40-infected cells is entirely cell-mediated, i.e., due to enzyme systems that are normally involved in recombination between cellular DNA segments. At present we understand too little about the function of SV40 gene products at the biochemical level to know whether SV40-coded proteins play a role. However, since SV40 has few genes, it seems likely that many, if not all, of the enzymes needed for recombination are cellular, and therefore, what is learned about SV40 recombination probably will have bearing on problems of recombination and gene mobility in normal animal cells.

58

Table 1. SV4O recombinant joints in
evolutionary variants (in SV4O map units)

Deletions	Duplications
.75/.11	.46/.69
.185/.525	.56/.72
.24/.54	.63/.78
.17/.74	.45/.76
.15/.72	
.72/.755	
.73/.99	
.165/.625	

Each pair of coordinates represents the
joint present in a cloned evolutionary
variant. See the legend of Figure 7a
for references

Acknowledgments. The authors' research reported in this publication
was supported by the U.S. National Cancer Institute (5 PO1 CA16519)
and the Whitehall Foundation, Inc.

References

Botchan, M., Topp, W., Sambrook, J.: Cell 9, 269-288 (1976)
Brockman, W.W., Gutai, M.W., Nathans, D.: Virology 66, 36-52 (1975)
Brockman, W.W., Lee, T.N.H., Nathans, D.: Virology 54, 384-397 (1973)
Brockman, W.W., Nathans, D.: Proc. Nat. Acad. Sci. USA 71, 942-946 (1974)
Chow, L.T., Boyer, H.W., Tischer, E.G., Goodman, H.M.: Cold Spring Harbor Symp.
    Quant, Biol. 39, 109-117 (1974)
Davoli, D., Ganem, D., Nussbaum, A.L., Fareed, G., Howley, P.M., Khoury, G., Martin,
    M.A.: Virology 77, 110-124 (1977)
Frenkel, N., Lavi, S., Winocour, E.: Virology 60, 9-2O (1974)
Hsu, M.-T., Jelinek, W.R.: Proc. Nat. Acad. Sci. USA 74, 1631-1634 (1977)
Kelly, T.J., Jr.: J. Virol. 15, 1267-1272 (1975)
Kelly, T.J., Jr., Lewis, A.M., Jr., Levine, A.S., Siegel, S.: J. Mol. Biol. 89,
    113-126 (1974)
Kelly, T.J., Jr., Nathans, D.: Adv. Virus Res. 21, 85-173 (1977)
Kelly, T.J., Jr., Rose, T.: Proc. Nat. Acad. Sci. USA 68, 1037-1041 (1971)
Ketner, G., Kelly, T.J., Jr.: Proc. Nat. Acad. Sci. USA 73, 1102-1106 (1976)
Lavi, S., Winocour, E.: J. Virol. 9, 309-316 (1972)
Lebowirt, P., Kelly, T.J., Jr., Nathans, D., Lee, T.N.H., Lewis, A.M., Jr.: Proc.
    Nat. Acad. Sci. USA 71, 441-445 (1974)
Lee, T.N.H., Brockman, W.W., Nathans, D.: Virology 66, 53-69 (1975)
Lewis, A.M., Jr., Breeden, J.H., Wewerka, Y.L., Schnipper, L.E., Levine, A.S.: Cold
    Spring Harbor Symp. Quant. Biol. 39, 651-656 (1974)
Mertz, J.E., Carbon, J., Herzberg, M., Davis, R.W., Berg, P.: Cold Spring Harbor
    Symp. Quant. Biol. 39, 69-84 (1974)
Morrow, J.F., Berg, P., Kelly, T.J., Jr., Lewis, A.M., Jr.: J. Virol. 12, 653-658 (1973)
Oren, M., Kuff, E.L., Winocour, E.: Virology 73, 419-430 (1976)
Subramanian, K.N., Dhar, R., Weissman, S.M.: J. Biol. Chem. 252, 333-339 (1977a)
Subramanian, K.N., Dhar, R., Weissman, S.M.: J. Biol. Chem. 252, 355-367 (1977b)
Rosenberg, M., Segal, S., Kuff, E.L., Singer, M.F.: Cell in press (1977)
Rozenblatt, S., Lavi, S., Singer, M.F., Winocour, E.: J. Virol. 12, 501-510 (1973)
Shen, C.-K.J., Hearst, J.E.: Proc. Nat. Acad. Sci. USA 74, 1363-1367 (1977)
Tai, H.T., Smith, C.A., Sharp, P.A., Vinograd, J.: J. Virol. 9, 317-325 (1972)
Uchida, S., Yoshiike, K., Watanabe, S., Furuno, A.: Virology 34, 1-8 (1968)

Adenovirus – Integration and Oncogenicity

J. K. McDougall, L. B. Chen, A. R. Dunn, and P. H. Gallimore

It is not within the purview of this article to give a complete review
of the properties of adenoviruses, which, although a defined and morphologically distinct virus group, exhibit a wide range of pathogenic
responses in the many species from which they have been isolated or in
which their oncogenic capabilities have been studied. Extensive reviews
on adenoviruses have been published recently (Wold et al., 1977; Philipson et al. 1976).

The adenovirus virion is icosahedral with an average diameter of 73 nm.
The capsid is composed of 252 capsomeres, each having a diameter of
8 nm. The twenty triangular faces of the icosahedron are formed by 240
capsomeres – hexons – and there are 12 capsomeres at the vertices-pentons.
Each penton has a rod-like projection described as a fiber. The three
structural proteins have been mapped on the genome (Flint, 1977). The
adenoviruses contain 11.6.-13.5% DNA (Pina and Green, 1965) as an un-
interrupted linear duplex of $20\text{-}25 \times 10^6$ daltons (Green et al., 1967;
Van der Eb et al., 1969) with inverted terminal repetitions that may
be recognition sites for a protein having the propensity to circularize double-stranded adenovirus DNA (Robinson et al., 1973).

Infection of permissive cells with adenovirus results in two distinct
phases described as "early" and "late," the transition from early to
late being coincident with and dependent upon the onset of viral DNA
synthesis. In nonpermissive cells this transition does not occur and
in the absence of viral DNA replication no late gene products, e.g.,
hexon, penton, and fiber structural proteins, are synthesized. The
result of such a host cell-mediated block is an abortive virus infection from which some cells survive, a proportion of these acquiring
neoplastic properties and new phenotypic characteristics at least some
of which are determined by the inducing virus.

Adenovirus-transformed Cells

In this paper we will concentrate on the characteristics of rodent and
human cells transformed by adenoviruses of human origin, serotypes 2,
5, and 12. The human adenoviruses can be classified into three main
groups on the basis of tumor induction in newborn hamsters. Type 12
is a member of group A, defined as highly oncogenic (Huebner, 1967),
types 2 and 5 are in group C and do not produce tumors on direct
inoculation of the virus into animals but will transform cells in vitro
at low frequency (McBride and Wiener, 1964; Freeman et al., 1967;
MacAllister et al., 1969). The highly oncogenic Ad12 virus is nonpermissive in all cell systems other than human, Ad2 and Ad5 transforma-
tions result from semipermissive infections of rodent cells (Gallimore,
1974; Williams, 1973). The transformed cell lines established after
Ad2 or Ad5 infection vary in their ability to induce tumors upon inoculation into syngeneic hosts or nude mice (Harwood and Gallimore, 1975).

Integrated Viral DNA in Adenovirus-transformed Cells

The examination of Ad2- and Ad5-transformed cell DNA for the presence of viral sequences, using ^{32}P-labeled restriction endonuclease fragments as probes (Gallimore et al., 1974; Sambrook et al., 1974; Flint et al., 1976), has identified in the case of each virus a region of the viral molecule ubiquitous in all cell lines. These viral DNA sequences represent 12-14% of the left-hand G+C-rich area of the genome, and some cell lines contain only this region in a few copies. None of the twelve lines examined contains sequences representing all areas of the virus genome, and the deletions are of unequal length and occur in different positions along the viral DNA molecule in different cell lines. It is difficult to postulate a role for sequences other than those common to all the Ad2- or Ad5-transformed cell lines (i.e., the 12-14%) if one considers only those characteristics associated with the virus transformed phenotype. A consistent RNA species, corresponding to a subset of those synthesized early in productive infection (Fujinaga et al., 1969; Green et al., 1970), is detectable in all the transformed cell lines (Flint et al., 1975; Flint and Sharp, 1976) and is transcribed from about half of the 12-14% left-hand region. Other regions of the viral DNA molecule may also be transcribed but there is no consistent pattern between different cell lines other than the fact that they correspond only to RNA synthesized "early" in productive infection. There is also evidence that with one exception the frequency of viral RNA is roughly proportional to the concentration of viral template integrated into host DNA (Flint and Sharp, 1976).

The integration pattern of Ad12 DNA in transformed cells differs from the Ad2 and Ad5 situation in that the majority, if not all, of the viral sequences are present in multiple copies (Fanning and Doerfler, 1976; Green et al., 1976; Galloway, personal communication). It has, however, been reported (Van der Eb, personal communication) that a fragment representing the left-hand G+C-rich 16% of the Ad12 genome is capable of transforming rat cells, analogous to the findings with group C viruses (Gallimore et al., 1974) and to the demonstration (Graham et al., 1974) that a virus DNA fragment of Ad2 from the extreme left-hand 8% of the viral genome will transform cells in vitro.

Localization of Viral Sequences

The mapping of DNA tumor virus genomes has progressed rapidly with the advent of restriction endonucleases, which produce defined fragments of viral DNA. The same technology has allowed a more precise analysis of the viral DNA fragments persisting in virus-transformed cells.

The next stage of analysis is to determine where in the host genome these viral sequences are integrated. If a specific integration site is necessary for stable transformation, are multiple copies integrated in tandem, or is one copy of the vital region of viral DNA sufficient at a specific site, with other copies persisting in "silent" noneffective loci or even as episomes? Also, where fragments of viral DNA are present at different frequencies, are these fragments distributed randomly or are they all present at one site?

Segregation of the chromosomes of adenovirus-transformed cells in interspecies somatic cell hybrids provides a method for examining the localization of viral sequences in the parental transformed cell line. Since adenovirus-transformed cells do not release infectious virus nor do they synthesize virus DNA (Dunn et al., 1973), they are ideal for studying virus-gene localization and the role these viral sequences play in initiating and maintaining transformation. This system is not influenced by the secondary interaction of inducible virus with the heterokaryon, as is the case in some SV40-transformed human cells. Studies of hybrid cell lines derived by the fusion of SV40-transformed human cells with mouse cells indicated integration sites for SV40 virus in human chromosomes 7 and 17 (Croce et al., 1973; McDougall et al., 1976; Croce, 1977) and have shown that presence of viral DNA, expression of viral T-antigen, and the transformed phenotype are all syntenic in these hybrids.

We have developed hybrid cell lines between adenovirus-transformed rat cells (Gallimore et al., 1974) and mouse cells and examined them for 1) identification of rat chromosomes and isozymes, 2) transcription of integrated viral genes using in situ hybridization to detect mRNA and 3) viral antigens by immunofluorescence.

Somatic cell hybrids were derived by fusion of virus-transformed rat cells and mouse cells deficient in thymidine kinase. Fusion was mediated by β-propiolactone-inactivated sendai virus, which is known to increase the frequency of cell fusion. A double selective system has been employed that favors the survival of hybrid cells over the parental cells. After fusion, cells were incubated in the presence of HAT (hypoxanthine, aminopterin, thymidine and glycine – Littlefield 1964), which selects against cells deficient in thymidine kinase (TK), and in growth medium containing calcium ions. Adenovirus-transformed rat cells, which are more sensitive to Ca^{2+} than normal cells (Freeman et al., 1967), tend to detach from the dish at levels above 0.5 mM $CaCl_2$. The surviving hybrid cells must necessarily contain the rat gene coding for thymidine kinase and are not Ca^{2+} sensitive. Fourteen to twenty-one days after seeding the mixed fused cell population at low density and incubating in the selective medium, morphologically distinct clones of cells were picked, plated out at low density, and re-cloned. A total of 57 cloned hybrid lines was established in which rat chromosomes had been selectively lost.

A method has been used that allows assignment of individual chromosomes to the species of origin. Mouse DNA contains a minor component of satellite DNA comprising 9-10% of the total DNA (Kit, 1961) and is located at the centromeres of the mouse chromosomes (Jones, 1970). Complementary RNA (cRNA) synthesized in vitro using mouse satellite DNA as a template has been hybridized in situ to fixed denatured chromosome preparations of both parental and hybrid cell lines.

Figures 1 and 2 are autoradiographs of metaphase chromosome spreads of hybrid cells. Figure 1 is a hybrid cell consisting of many rat and mouse chromosomes, unlike the cell shown in Figure 2, which contains only 3 rat chromosomes and is representative of hybrid Ad2/F4/LMTK⁻/B1, which is a T-antigen positive cloned line derived from the original mixed hybrid cell population. The 3 rat chromosomes present in this line have been examined using Giemsa banding and have been identified as chromosomes 2, 17, and 19, as shown, inset in Figure 2. In the hybrid lines examined, the only rat chromosomes positively identified are numbers 1, 2, 11, 12, 14, 17, and 19; however, none of these chromosomes is consistently present in every hybrid cell line. In the majority of cloned cell lines, the content of rat chromosomes remains stable. Since all the hybrid cell lines survive in HAT medium, they must by definition contain the rat thymidine kinase (TK) gene.

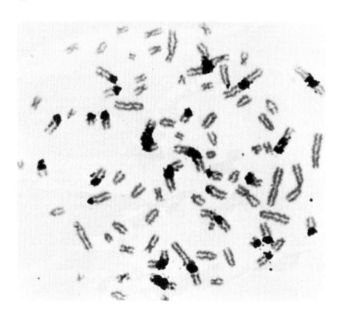

Fig. 1. In situ hybridization of a denatured (0.07N NaOH for 3 min at 22°C) preparation of hybrid cells. Fifty thousand cpm of mouse satellite cRNA (estimated spec.
act. 10^7 cpm/µg) in 2.5 × SSC was hybridized for 20 min at 58°C after which slides
were treated with RNAse, 20 µg/ml in 2 × SSC for 30 min at 37°C, and washed extensively in 2 × SSC at 4°C. Autoradiographs were exposed for 7 days at 4°C and
stained in 10% Giemsa, pH 6.8. This cell represents an LMTK⁻/Ad2 hybrid cell from
the first subculture of an uncloned cell population. Autoradiographic grains represent hybridization of mouse satellite cRNA to mouse centromeres. Cells of this type
with a predominance of rat chromosomes do not give rise to stable cell lines

The localization of integrated viral sequences in transformed cell
chromosomes presents a challenge of greater complexity than that associated with the mapping and linkage analysis of host genes. The finding that adenovirus or SV40 DNA molecules can persist as multiple copies
of fragments occurring at different frequencies does not seem to be
immediately compatible with the concept of integration at a single
locus. It is nevertheless possible that one integration site could
have more significance that others, and the initial results with SV40-
transformed human and primate cells were consistent with this hypothesis. The significance of such a locus would be 1) that it allows
transcription of the integrated viral genes by utilization of host
promoters, resulting in a product controlling the transformed phenotype or 2) that the integrated viral DNA effects a steric change in
the host genome with consequent alteration in host transcription and
regulation. There is evidence to suggest that the number of integration
sites may be limited (Prasad et al., 1975).

The original experiments described by Croce and his colleagues (Croce
et al., 1973; Croce and Koprowski, 1974; Khoury and Croce, 1975) showed
remarkable consistency in retention of human chromosome C7 and the SV40-
transformed phenotype in somatic cell hybrids. This result may suggest
a virus integration site of primary importance in this autosome, but
it does not rule out the possibility of integration sites in other
chromosomes. Preferential retention of a chromosome could occur for
reasons other than the content of SV40 genes, one being that there

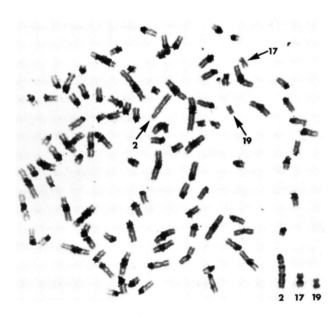

Fig. 2. In situ hybridization of mouse satellite cRNA with the Ad2/F4/LMTK⁻/B1 hybrid cell line. Hybridization was carried out as described in legend to Figure 1. This line contains 3 identifiable rat chromosomes, i.e., Nos. 2, 17, and 19, and the banded preparations of these chromosomes are shown in the inset of this Figure. Rat chromosomes defined according to Committee for a Standardized Karyotype of *Rattus norvegicus*. Cytogenet. Cell Genet. (1973) 12, 199

might be a growth advantage conferred upon the hybrid cell by host genes located on that chromosome. There is indeed some evidence for preferential retention of C7 in hybrids where SV40 is not implicated (Kucherlapati and Ruddle, 1975). A preexisting, though minimal, growth advantage for hybrids containing C7 or 17 may well be amplified by integrated SV40 genes, which by definition already confer growth and survival advantages on cells. Such a process does not preclude the stable integration of SV40 DNA into other chromosomes of the parental virus-transformed cell.

Our studies with adenovirus-transformed cells have provided a series of hybrids, some of which contain virus DNA sequences that serve as templates for virus RNA transcription. Although our data show that certain rat chromosomes appear with high frequency in hybrid cell lines, some of which contain virus mRNA and express T-antigen, the integration site(s) of adenovirus DNA sequences in this limited number of transformed cell lines does not appear to have segregated together with a single identifiable rat chromosome or isozyme.

Attempts to detect integrated adenovirus DNA sequences, in transformed cells and virus-induced tumors, using the in situ hybridization method with ³H-cRNA as a probe, were reported previously (Dunn et al., 1973; Loni and Green, 1973). Although the methods used would not have been sufficiently sensitive to detect the minimal amounts of viral DNA integrated into some cell lines, e.g., Ad2/F17 C8 contains approximately 11 × 10⁶ daltons of Ad2 DNA per diploid quantity of cell DNA, the multiple copies found in Ad2/F4 and in Ad12 tumor cells are above the calculated minimum sensitivity of the technique. The results of

these experiments indicated an association of autoradiographic label
with many chromosomes and not with any specific sites on the chromosomes.

Adenovirus-transformed cell lines are karyotypically abnormal and the
parental rat cell lines used to produce these hybrids are no exception
to the rule. Ad2/F4 has a majority of cells that are triploid and has
metaphases ranging from hypotriploid to hypertriploid. There are 2-3
abnormal and stable marker chromosomes in every cell and one chromosome (No. 2) is consistently involved in the genesis of markers in
both Ad2/F4 and Ad2/F17 C8. The fact that none of the hybrid cell lines
that are positive for viral mRNA and T-antigen contains any marker
chromosome from the virus-transformed cells indicates that the viral
integration site is not associated with the markers. In a recent study
of herpesvirus type 2-transformed hamster cells, no correlation was
found between marker chromosomes, viral sequences, and oncogenicity
(Copple and McDougall, 1976).

If homologous chromosomes have the same capability to integrate viral
sequences, as might be expected, then the triploidy of these rat lines
would tend to result in a dilution of viral gene dosage in hybrid
lines when, as we have found, there is a tendency for only haploid
copies of rat chromosomes to be retained in the hybrids. This is compatible with our observation that there is decreased T-antigen immunofluorescence in the positive hybrid cells, compared with the transformed rat cell. This is particularly true of Ad2/F17 C8, which by C_ot
analysis has only 3.5 copies of the left-hand 14% of the viral genome
per diploid quantity of cell DNA.

The only well-characterized adenovirus-transformed human cell line
(Graham et al., 1974) is similar. in that these cells again tend to
be heterogeneous in chromosome constitution with an average count in
the triploid range and with only a small amount of viral DNA per diploid
quantity of human DNA (Sambrook, personal communication). Our preliminary experiments with this cell line hybridized to TK$^-$ mouse cells
have again indicated a dilution of T-antigen and an association of this
viral function with retention of a human chromosome fragment (Fig. 3).

The methods described allow selection of hybrid lines for a more critical quantitative analysis of viral sequences and the part these sequences play in determining the transformed and oncogenic phenotypes.
This approach and that described by Botchan et al. (1976) together
with improved in situ hybridization techniques (Moar and Jones, 1975)
should eventually provide a better understanding of the nature of integration of virus genes in transformed cells. The somatic cell hybrid
approach is the optimum method for determining in any given cell line
whether 1) the integration of specific viral genes, 2) expression of
T-antigen, 3) the transformed phenotype in vitro and 4) oncogenicity
in vivo are related phenomena that cannot be segregated.

Oncogenicity, LETS Protein, and Integrated Viral DNA

Recently, we turned our attention to the intriguing observation that
there are several adenovirus-transformed cell lines that are nontumorigenic (Harwood and Gallimore, 1975). For instance, cell lines Ad2/F17
and Ad2/F18, which are rat embryo cells transformed by adenovirus type
2, are T-antigen positive and able to grow in low serum and to high
saturation density; yet, they are nontumorigenic in normal syngeneic

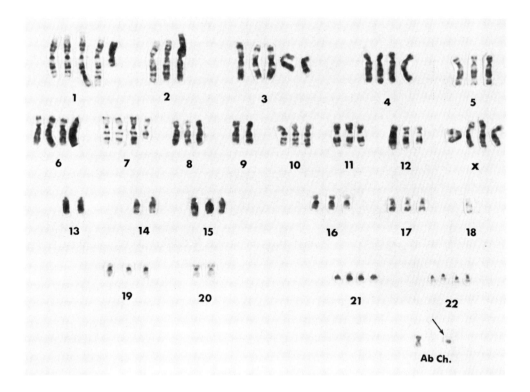

Fig. 3. Giemsa-banded chromosomes of the adenovirus 5 DNA-transformed human embryo kidney cell line (Ad5/HEK FG 293-31). Segregation of the chromosomes in somatic cell hybrids has given adenovirus T-antigen positive hybrids only when the unidentified abnormal chromosome fragment (arrowed) is present

rats, immunosuppressed newborn syngeneic rats, and nude mice. Ad2/F19, on the other hand, is tumorigenic in nude mice, but not in rats, whether immunosuppressed or not. In comparison with the other Ad2 lines studied, Ad2/F19 showed a lower level of tumor induction in nude mice. Ad2/F19 tumors had a longer latent period than the other lines (e.g., T2C4, 7.5 days; REM, 8.5 days; F4, 19 days; F19, 30 days): whereas T2C4, REM, and F4 showed local invasion of mouse tissues, this was not the case with Ad2/F19, which classified histologically as a benign tumor. Ad2/F4 and Ad2/REM are tumorigenic in immunosuppressed syngeneic rats while some other lines (e.g., T2C4) are tumorigenic in syngeneic rats without immunosuppression. This series of cell lines thus provides a spectrum of oncogenicity within a single species.

Recently, a cell surface iodinated protein with nominal molecular weight of about 250,000 was shown to be either undetectable or reduced in various viral-transformed fibroblasts. This protein has been designated as LETS (large external protein that is transformation sensitive) protein (Hynes and Bye, 1974). When normal rat embryo fibroblasts were examined for the distribution of LETS protein by immunofluorescence, the pattern found was density and time dependent. In a very sparse culture, most of the surface LETS proteins are located at cell-substratum contact areas in a fine fibril-like structure. In a monolayer culture grown for 48 h, surface LETS protein is distributed in a diffuse network over the cell surface of all cells. After 6 days, the

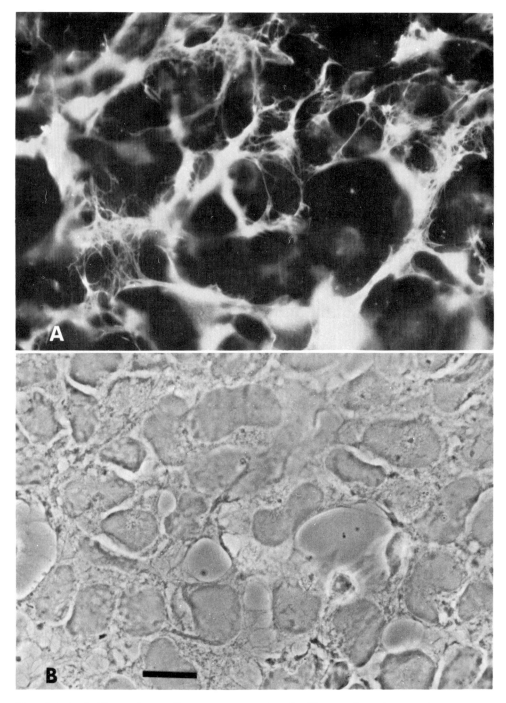

Fig. 4 A-C. Indirect immunofluorescence stain of Ad2-transformed rat cells. Method
as described in Chen et al. (1976) Phase contrast (B) and immunofluorescent (A)
micrographs of same area of Ad2/F19 monolayer showing extensive, but reduced from
normal, fibrillar network of LETS protein. This line is nontumorigenic in rats and

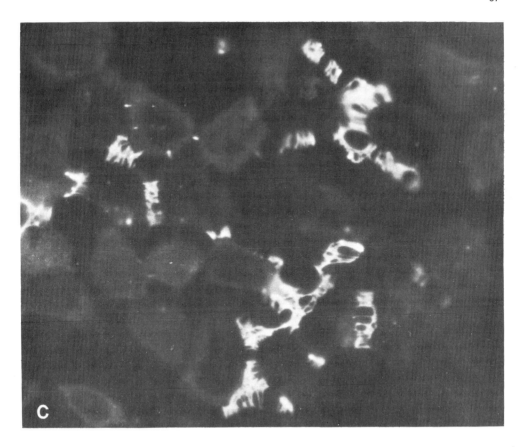

Fig. 4 (continued)
only minimally tumorigenic in nude mice. (C) LETS protein in cell-cell contact areas
only in cell line Ad2/B1. This line is tumorigenic in immunosuppressed rats and in
nude mice

culture is covered with a massive network of fibril-like structures
composed of LETS protein.

As shown in Figure 4, when the distribution of surface LETS protein
in a series of adenovirus-transformed rat cells was studied by indirect
immunofluorescence, fibril-like antigens, when detected, were always
located in the contact area between cells. When cells are not in contact, LETS protein is rarely detected in this series of adenovirus-
transformed cells. When the number of cells positive for LETS protein
upon contact was scored in these lines, a correlation between decrease
in the cells positive for LETS protein and increase in tumorigenicity
is observed (Chen et al., 1976). When other cells with known tumorigenicity were also grown in monolayer and assayed for LETS protein
on the cell surface upon contact, the correlation was equally valid.

We wish to emphasize that the induction of tumors in animals by viral-
tranformed cells prepared in culture must result from alterations in a
series of cellular properties. Although the cell surface will play a
crucial role, it is by no means the sole or the prime factor involved

in oncogenicity. Moreover, even in the domain of cell surface alteration surface LETS protein is unlikely to be the only change relevant
to oncogenicity. It is expected that other factors including host
response and, for example, angiogenesis-stimulating factor (to assure
a nutrient supply), hydrolytic enzymes (for tissue invasiveness),
alterations in the pattern of cell-hormone interactions (for autono-
mous cell growth), and the whole immune system will all play an important role in the determination of oncogenic potential of transformed
cells.

We do not know why there should be a correlation between the loss of
LETS protein and tumor induction. The simplest explanation may be
that LETS protein is directly involved in growth control and that the
loss of LETS protein is responsible for unrestricted growth during
tumor formation. Our previous finding (Teng and Chen, 1975) argues
against this possibility. Perhaps the maintenance of a normal pattern
of cell growth in vivo depends on a proper intercellular matrix system.
The integrity of such a matrix may be impaired by the loss of one
(LETS protein) or two (LETS protein and, for example, collagen) of
its elements. An important observation is that of all the adenovirus-
transformed cell lines described, only Ad2/F17, F18, and F19 produce
a three-dimensional matrix of LETS protein on the cell surface after
a longer (6-day) period in culture, similar to that seen with normal
cells. Results from further experiments indicate that loss of capacity
to form such an exoskeleton may be of prime importance in oncogenic
behavior.

From the data available on the Ad2 transformed cell lines (Gallimore
et al., 1974), there is no clear relationship between content of integrated viral sequences and oncogenicity or between viral sequences and
loss of LETS protein. One consistent factor that can be correlated
with oncogenicity of Ad2-transformed cells is the amount of infectious
virus used to infect the monolayer cultures in which transformed foci
develop (McDougall, 1975). Cells transformed at a low m.o.i., e.g.,
0.25 plaque forming units (p.f.u.) per cell, are either nononcogenic
even in immunosuppressed animals or nude mice or will only produce
benign tumors that are encapsulated, arise after a long latent period,
and do not transplant to syngeneic newborn rats (Gallimore et al.,
1977). With increased virus m.o.i. (e.g., 1-10 pfu) cell lines are
derived that induce transplantable tumors in immunosuppressed syngeneic
rats and at a m.o.i. of 50 pfu/cell, the virus-transformed cell lines
are completely oncogenic without any requirement for immunosuppression.

There is some indication (Sambrook et al., 1974) that cells transformed
at higher m.o.i. retain more of the virus genome, but an exception to
this is the Ad2/B1 cell line and the tumor lines B1/T4 and B1/T8 derived from it; these cell lines contain only the 14% left-hand end of
the virus genome and are oncogenic. It may be significant that examination of cells transformed by the highly oncogenic serotype Ad12 has
resulted in identification integrated sequences representing most or
all of the virus genome. This may mean that cells retaining more of
the virus DNA sequences are more likely to be oncogenic and that all
Ad12 sequences are readily retained by the host genome.

Acknowledgements. We thank Dr. James D. Watson for advice and support
and Ms. Madeline Szadkowski for preparation of the manuscript. This
work is supported by the National Cancer Institute, USA, the Cancer
Research Campaign, England, and the Muscular Dystrophy Association of
America.

References

Botchan, M., Topp, W., Sambrook, J.: Cell 9, 269-287 (1976)
Chen, L.B., Gallimore, P.H., McDougall, J.K.: Proc. Natl. Acad. Sci. USA 73, 3570-3574 (1976)
Copple, C.D., McDougall, J.K.: Int. J. Cancer 17, 501-510 (1976)
Croce, C.M.: Proc. Natl. Acad. Sci. USA 74, 315-318 (1977)
Croce, C.M., Girardi, A.J., Koprowski, H.: Proc. Natl. Acad. Sci. USA 70, 3617-3620 (1973)
Croce, C.M., Koprowski, H.: J. Exp. Med. 140, 1221-1229 (1974)
Dunn, A.R., Gallimore, P.H., Jones, K.W., McDougall, J.K.: Int. J. Cancer 11, 628-636 (1973)
Fanning, E., Doerfler, W.: J. Virol. 20, 373-383 (1976)
Flint, S.J.: Cell 10, 153-156 (1977)
Flint, S.J., Sharp, P.A.: J. Mol. Biol. 106, 749-771 (1976)
Flint, S.J., Gallimore, P.H., Sharp, P.A.: J. Mol. Biol. 96, 47-68 (1975)
Flint, S.J., Sambrook, J., Williams, J., Sharp, P.A.: Virology 72, 456-470 (1976)
Freeman, A.E., Black, P.H., Vanderpool, E.A., Henry, P.H., Austin, J.B., Huebner, R.J.: Proc. Natl. Acad. Sci. USA 58, 1205-1212 (1967)
Fujinaga, K., Pina, M., Green, M.: Proc. Natl. Acad. Sci. USA 64, 255-262 (1969)
Gallimore, P.H.: J. Gen. Virol. 25, 263-273 (1974)
Gallimore, P.H., McDougall, J.K., Chen, L.B.: Cell, 10, 669-678 (1977)
Gallimore, P.H., Sharp, P.A., Sambrook, J.: J. Mol. Biol. 89, 49-72 (1974)
Graham, F.L., Abrahams, P.J., Warnaar, S.O., Mulder, C., De Vries, F.A.J., Fiers, W., Van der Eb, A.J.: Cold Spring Harbor Symp. Quant. Biol. 39, 637-650 (1974)
Green, M., Chinnadurai, G., Mackey, J.K., Green, M.: Cell 7, 419-428 (1976)
Green, M., Parsons, J.T., Pina, M., Fujinaga, K., Caffier, H., Landgraf-Leurs, I.: Cold Spring Harbor Symp. Quant. Biol. 35, 803-818 (1970)
Green, M., Pina, M., Kimes, R.C., Wensink, P.C., MacHattie, L.A., Thomas, C.A.: Proc. Natl. Acad. Sci. USA 57, 1302-1309 (1967)
Harwood, L.M.J., Gallimore, P.H.: Int. J. Cancer 16, 498-508 (1975)
Huebner, R.J.: In: Perspect. Virol. Pollard, M. (ed.). New York: Academic Press, 1967, Vol. V, pp. 147-166
Hynes, R.O., Bye, J.M.: Cell 3, 113-120 (1974)
Jones, K.W.: Nature (London) 225, 912-915 (1970)
Khoury, G., Croce, C.M.: Cell 6, 535-542 (1975)
Kit, S.: J. Mol. Biol. 3, 711-716 (1961)
Kucherlapati, R.S., Ruddle, F.H.: Ann. Intern. Med. 83, 553-560 (1975)
Littlefield, J.W.: Science 145, 709-710 (1964)
Loni, M., Green, M.: Virology 12, 1288-1292 (1973)
McAllister, R.M., Nicolson, M.O., Lewis, A.M., MacPherson, I., Huebner, R.J.: J. Gen. Virol. 4, 29-36 (1969)
McBride, W.D., Wiener, A.: Proc. Soc. Exp. Biol. Med. 115, 870-874 (1964)
McDougall, J.K.: In: Prog. in Med. Virology. Melnick, J.L. (ed.). Karger: Basel, 1975, Vol. 21, pp. 118-132
McDougall, J.K., Gallimore, P.H., Dunn, A.R., Webb, T.P., Kucherlapati, R.S., Nichols, E.A., Ruddle, F.H.: In: Third Intern. Conf. on Human Gene Mapping. The National Foundation: New York, 1976, Birth Defects: Original Article Series XII, 7, pp. 206-210
Moar, M.H., Jones, K.W.: Int. J. Cancer 16, 998-1007 (1975)
Phillipson, L., Pettersson, U., Lindberg, U.: In: Virol. Monogr. Gard, S., Hallauer, C. (eds.). New York: Springer-Verlag, 1975, Vol. 14
Pina, M., Green, M.: Proc. Natl. Acad. Sci. USA 54, 547-551 (1965)
Prasad, I., Zouzias, D., Basilico, C.: J. Virol. 16, 897-904 (1975)
Robinson, A.J., Younghusband, H.B., Bellett, A.J.D.: Virology 56, 54-69 (1973)
Sambrook, J., Botchan, M., Gallimore, P.H., Ozanne, B., Pettersson, U., Williams, J., Sharp, P.A.: Cold Spring Harbor Symp. Quant. Biol. 39, 615-632 (1974)

Teng, N.N.H., Chen, L.B.: Proc. Natl. Acad. Sci. USA 72, 413-417 (1975)
Van der Eb, A.J., Kestern, L.W., Van Bruggen, E.F.J.: Biochim. Biophys. Acta 182, 530-541 (1969)
Williams, J.F.: Nature (London) 243, 162-163 (1973)
Wold, W.S.M., Green, M., Buttner, W.: In: The Molecular Biology of Animal Viruses. Nayak, D.P. (ed.). New York: Marcel Dekker, 1977

The Presence and Expression of *Herpes Simplex* Virus DNA in Transformed Cells

N. Frenkel and J. Leiden

Introduction

Studies of papova and adenoviruses have led to the identification of
the viral DNA sequences which are involved in the transformation of
cultured cells by these groups of viruses (1-6, and reviewed in 7).
These viral DNA sequences have been shown to be stably incorporated
into the transformed cell genome (8-9). Their presence in transformed
cells has been correlated with their transcription into RNA (10-13),
as well as with the expression of identifiable, viral specific, transformation antigens (reviewed in 7, 14, 15). More recent research has
been aimed at elucidating the nature of the specific host and viral
DNA sites involved in the integration event (16-19).

Studies designed to identify the transforming region of *herpes simplex*
virus (HSV), and to characterize the nature of the association between
viral and host DNAs have only recently been initiated. In this paper,
we hope to present a critical discussion of these studies, as they
relate to the current, rather limited, state of knowledge concerning
HSV mediated transformation of cultured cells. For the sake of clarity,
this discussion will be divided into three parts as follows. First, we
will describe the various methods and cell systems which have been
used to transform cells in culture using HSV. Second, we will deal
with the expression and regulation of HSV genes in the transformed
cells. Finally, we will discuss current hybridization data related to
the presence of HSV nucleic acids in the various transformed cell lines
which have been investigated to date.

Transformation of Cultured Cells Using *Herpes Simplex* Viruses

Transformation of cells in culture by *herpes simplex* viruses has been
reported for cells from a variety of hosts including hamster, rat, and
human (20-30). Common to all reports is the necessity for inactivation
of the lytic potential of HSV prior to transformation. This inactivation has been accomplished by using UV irradiation or photodynamic inactivation (20, 21, 24, 27, 29, 30). Alternatively, infection under
nonpermissive conditions has been used to transform cells. For example,
transformation of cells with temperature-sensitive mutants of HSV at
nonpermissive temperatures (26-28) and with wild type virus, at supra-
optimal temperatures (22, 23, 27, 30), have been reported.

Cells transformed by HSV have been selected in two ways. The first
method, which has been referred to as biochemical transformation, involves the transfer of the HSV encoded thymidine-kinase (tk) gene to
cells previously lacking this enzyme (31-33). Such HSV-tk+ transformants
continue to express the viral specified thymidine kinase when selected
for and maintained in HAT medium. The relevance of this biochemical
transformation to oncogenicity, as defined by the ability to form

tumors when injected into the homologous host, has not yet been determined. The second type of assay which has been used to generate HSV transformed cell lines involves the selection of cells exhibiting altered growth properties, as reflected by their ability to form foci in culture (20-30). It has been reported that cell lines selected in this way exhibit variable degrees of tumorigenicity (35). This finding might be explained by the observation of Shin et al. that focus formation in culture correlates imperfectly with oncogenic potential in vivo (34).

Regulation and Expression of HSV Functions in Transformed Cells

Three types of studies have dealt with the expression of HSV specified functions in transformed cells. First, MacNab and Timbury (36) and Kimura et al. (37) reported the complementation of certain HSV temperature-sensitive mutants during infection of HSV transformed rat and hamster cells at nonpermissive temperatures. This complementation was presumably mediated by functionally expressed, resident HSV genes. It is interesting that the transformed cells were only able to complement a small subset of the mutants used in each study, suggesting selective expression of HSV functions in these cell lines. In a second set of studies, Lin and Munyon (38) and Leiden et al. (39) reported that the HSV thymidine kinase gene, present in mouse cells, which had previously been biochemically transformed to the tk+ phenotype with UV-irradiated HSV-1, remains responsive to the HSV products expressed during superinfection with a tk$^-$ virus. It remains to be determined whether the expression of the viral tk gene in the unsuperinfected tk$^+$ transformed cells is under viral or cell control. Finally, several laboratories have reported the presence of viral specified antigens in hamster, rat and human HSV transformed cells (20, 24, 26-30, 40-42). In each case viral antigens were detected only in a small proportion of cells in a given transformed cell culture. Two points are noteworthy with respect to these studies. First, the search for viral specific, transformation related, antigens has been complicated by the fact that HSV codes for approximately fifty polypeptides (43). Until recently, high titer, monospecific antisera were not available. The importance of using monospecific antisera in tests of transformed cells has recently been demonstrated by Flannery et al. (42). These authors showed that the use of monospecific antisera allowed the identification of a cystoplasmic and perinuclear HSV specific antigen (VP143) in four independently transformed hamster cell lines which were found to display only membrane flourescence when examined with polyspecific antisera. In light of this study it is fair to conclude that the identification of HSV specified transformation antigens awaits larger, comparative studies, using monospecific antisera. A second complication in the search for HSV specific transformation antigens was also suggested by Flannery et al. (42) who showed that the expression of viral antigen VP143 is cell-cycle dependent. Only by using synchronized cell cultures were these authors able to demonstrate strongly positive HSV specific immunofluorescence in 90% of the cells in a given culture. These two observations of Flannery et al. (42) may be important in explaining why most authors, using polyspecific antisera, have only observed HSV specified antigens in a small minority of the transformed cells in unsynchronized cell cultures.

HSV Nucleic Acids in Transformed Cells

Transcription of HSV DNA in Transformed Cells

Three studies have dealt with the presence of HSV specified RNA in transformed cells. First, by hybridizing labeled RNA, extracted from the HSV transformed hamster cell line 333-8-9, to HSV-2 DNA, immobilized on nitrocellulose filters, Collard et al. showed that a small portion of viral DNA sequences are transcribed in these cells (44). Second, Copple and McDougall utilized in situ hybridization with labeled HSV-2 DNA, and were able to detect the presence of viral RNA in an uncloned population, and in eight out of ten clones of the cell line 333-8-9 (45). Finally, we have shown that a limited portion of HSV-2 DNA was transcribed in two hamster transformed cell lines (MS4-1 and 333-2-29) tested by liquid hybridization of in vitro labeled HSV-2 DNA to RNA extracted from the transformed cells (N. Frenkel, unpublished data).

The Presence of Viral DNA Sequences in HSV Transformed Cells

A number of investigators have reported attempts to detect viral DNA sequences in cultured cells, transformed by HSV. As described above, two types of transformed cell lines have been studied: (1) mouse tk^+ cells, transformed by UV light irradiated HSV, and selected for their ability to express the viral specified thymidine kinase, and (2) hamster embryo cells, transformed by UV light or photoinactivated HSV-2 and selected for their ability to form foci. Each of the studies described below involved the analysis of the rate of reassociation of a labeled viral DNA probe in the presence of unlabeled test DNA, extracted from transformed cell lines. The data from these studies are summarized in Table 1, and are listed below:

Table 1. Summary of hybridization studies of HSV-1 and HSV-2 transformed cell lines

Authors	Probe DNA	Transformed cell line and tumor	Fraction of viral DNA retained	No. of copies per cell	$N =$ Cell DNA equiv. / Viral probe DNA equiv.
Kraiselburd et al. (1976)	HSV-1	LVtk+ clone 139	0.23	5	16.6
	HSV-1	LVtk⁻ clone 139	-0-	---	16.6
Davis and Kingsbury (1976)	HSV-1	LVtk+ clone 139	0.09±0.03	5	0.8
	HSV-2	LVtk+ clone 207	0.10±0.02	6	1.3
	HSV-2	333-8-9	-0-	---	1.3
Frenkel et al. (1976)	HSV-2	333-2-29 line A	0.37±0.02	3.1	90.8
	HSV-2	333-2-29 line B	0.10±0.04	1.5	39.6
	HSV-2	333-8-9 P80	0.16±0.02	0.7	53.1
	HSV-2	333-8-9 tumor	0.08±0.02	3.0	71.1
	HSV-2	333-8-9 T1-P32	0.09±0.01	ND	53.5
	HSV-2	Col 6-1	0.14±0.04	1.8	31.4
	HSV-2	MS 4-1	0.09±0.02	1.9	73.0
	HSV-2	333-2-26	0.10±0.03	2.4	26.7
Minson et al. (1976)	HSV-2	333-8-9	0.40	---	0.4
	HSV-2	333-8-9 clone	-0-	---	0.4

1. Biochemically transformed mouse tk+ cells: Kraiselburd et al. (46)
reported the presence of 23% of HSV-1 DNA sequences, in approximately
five copies per cell, in the mouse cells transformed by Munyon et al.
(31). Davis and Kingsbury studied the same mouse tk+ cells, and reported the presence of 4-6 copies per cell of 10% of HSV DNA sequences (47).

2. HSV transformed hamster cell lines selected by focus formation:
Davis and Kingsbury were unable to detect viral DNA sequences in the
hamster cell line 333-8-9 (47), produced by Duff and Rapp (20). Minson
et al. reported that HSV-2 DNA sequences were present in an early
passage of the cell line 333-8-9, but were not detectable in cells
from later passages of the same line, or in 17 clones established from
that line (48). As the authors point out, it should be noted that some
of these same clones revealed the presence of viral RNA in the in situ
hybridizations of Copple and McDougall described above (45). We reported the results of analyses of the hamster cell line 333-8-9, four
additional hamster cell lines produced by Duff and Rapp (20, 21, 25),
a tumor produced by innoculation of 333-8-9 cells into a newborn hamster, and a cell line derived from a second 333-8-9 induced tumor. As
can be seen in Figure 1, we were able to detect HSV-2 DNA sequences in
all cell lines tested. The fraction of viral DNA sequences represented
in the various cell lines varied from 8 to 37%, present in one to three
copies per cell (49). Parallel passages of one of these cell lines
(333-2-29) were found to contain different amounts of viral DNA sequences. The HSV-2 DNA sequences present in the hamster tumor, and in
the cell line derived from a second tumor had a lower sequence complexity than those present in the 333-8-9 cell line which was originally used to produce the tumor.

There are two possible explanations for the failure of some investigators to detect viral DNA in the transformed cells. First, it is possible that the maintenance of the transformed state may not require
the constant presence of viral genes in the transformed cells. Although,
at this time, we cannot exclude or confirm this "hit and run" hypothesis, it should be noted that evidence from the smaller oncogenic DNA
viruses, papovaviruses and adenoviruses, suggests that the retention
and expression of specific portions of the transforming viral genome
is necessary for the maintenance of the transformed state (50-54).
Alternatively, it is possible that some investigators failed to detect
viral DNA sequences in HSV transformed cells because their hybridization tests were not sufficiently sensitive to allow detection of a
small portion of the viral genome present in a small number of copies
per cell. Because the sensitivity of reassociation tests of the type
described above is greatly influenced by the relative amounts of the
viral DNA probe and the transformed cell DNA in the hybridization
mixture (55, and Frenkel, Cox and Roizman, submitted for publication),
it is of interest to examine the experimental details of the various
hybridization tests employed in the studies described above.

For comparative purposes we shall define a term N as the ratio of cell
DNA equivalents to viral probe DNA equivalents in the hybridization
mixture (i.e., moles of unlabeled cell DNA per mole of labeled viral probe
DNA). The N values used in the various studies described above are
listed in Table 1. A brief examination of that table reveals that viral
DNA sequences were always detected in tests employing high N values
(larger than 20). Parenthetically, we would like to add that, in our
own studies, involving the reassociation of labeled HSV-2 DNA in the
presence of unlabeled DNA extracted from various transformed hamster
cell lines, we were able to detect viral DNA sequences in tests constructed to yield N values of 20 or more, but not in tests constructed
to yield N values close to 1 (see Fig. 1). Furthermore, theoretical

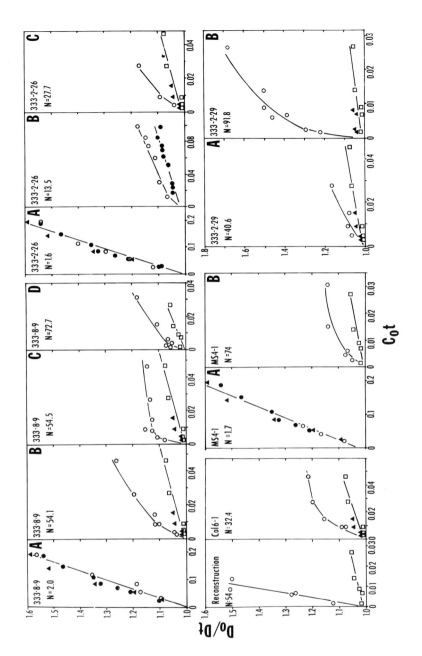

Fig. 1. Hybridization of HSV-2 DNA with hamster transformed cell lines and with a hamster tumor. Data from Frenkel et al. (49). ^{3}H-labeled (low N values) or ^{32}P-labeled (high N values) HSV-2 DNAs were reassociated in the presence of calf thymus (●), normal hamster (▲), and HEp-2 (□) DNA, or in the presence of DNA extracted from the various transformed cell lines (○). N is defined as the molar ratio of transformed cell DNA to viral probe DNA in the hybridization reactions. For example, 10 μg of cell DNA plus 10^{-4} μg of viral probe DNA corresponds to $N=1$, assuming molecular weights of 10^{13} and 10^8 daltons for the cellular and viral DNAs respectively

analysis of the hybridization kinetics shows that we cannot expect to detect sequences corresponding to less than 5% of the probe DNA, even in tests constructed to yield high N values (55). Thus, on theoretical grounds, viral DNA sequences corresponding to as much as 5×10^6 daltons in complexity (i.e., 2.5 - 3-fold the estimated size of the transforming regions of papova and adenoviruses) would have escaped detection in tests involving intact HSV DNA as the labeled probe.

In summarizing the available hybridization data, we would like to end this review with a discussion of several, tentative conclusions which can be made concerning the presence and retention of HSV DNA sequences in transformed cells. First, as summarized in Table 1, and discussed above, tests performed at high N values (i.e., high ratios of transformed cell DNA to viral DNA probe) detected HSV DNA sequences in all of the cell lines studied. Second, only a portion of the viral genome was detected in each cell line. Third, in all cases, only a relatively small number of copies per cell of viral sequences was found. Finally, it is clear from all studies reported that there is marked heterogeneity between the fractions of the viral genome retained in the various transformed cell lines. This heterogeneity can be seen to occur both from one cell line to another, and between different clones of the same transformed line, as well as after passaging of a cell line, either in culture, or through animals.

There are several possible explanations for the heterogeneity which has been observed. First, as mentioned previously, it is possible that the continued presence and expression of HSV genes in the transformed cells is not necessary for the maintenance of the transformed phenotype. Second, it is possible that the original transformed cell population was composed of subpopulations of cells, containing different amounts of viral DNA sequences. We would expect such heterogeneity either if the original cell line was derived from more than one independent transformation event, or if HSV DNA sequences which are nonessential for the maintenance of the transformed phenotype are spontaneously lost from subpopulations of these transformed cells. Once such heterogeneity exists within a transformed cell population, it is not unreasonable to expect that passaging in culture, or cloning, could result in the selection of cells retaining smaller amounts of viral DNA, or containing viral DNA sequences which cannot be expressed. Such reasoning follows from the fact that HSV is a highly lytic virus, with potent inhibitory effects on host macromolecular synthesis. The stability of a given viral DNA sequence in the transformed cells might depend upon whether or not it is covalently linked to the host DNA and, if so, upon the specific viral and host sites of integration. Experiments currently underway, using restricted portions of the HSV genome as probes in highly sensitive hybridization assays, should shed much light on both the identity of the HSV transforming genes, and on the nature of their association with the host genome.

Acknowledgments. The studies conducted at The University of Chicago were aided by grants from the American Cancer Society (VC204), the National Cancer Institute USPHS (CA-19264) and the Leukemia Research Foundation. Jeffrey Leiden is a predoctoral trainee (USPHS 5-T32 HD-07009-03).

References

1. Graham, F.L., Van der Eb, A.J., Heijneker, H.L.: Nature (New Biol.) 251, 687 (1974)
2. Graham, F.L., Abrahams, P.J., Mulder, C., Heijneker, H.L., Warnaar, F.A., De Vries, F.A.J., Fiers, W., Van der Eb, A.J.: Cold Spring Harbor Symp. Quant. Biol. 39, 637 (1974)
3. Sambrook, J., Botchan, M., Gallimore, P., Ozanne, B., Pettersson, U., Williams, J., Sharp, P.A.: Cold Spring Harbor Symp. Quant. Biol. 39, 615 (1974)
4. Sharp, P.A., Pettersson, U., Sambrook, J.: J. Mol. Biol. 86, 709 (1974)

5. Gallimore, P.H., Sharp, P.A., Sambrook, J.: J. Mol. Biol. 89, 49 (1974)
6. Botchan, M.B., Ozanne, B., Sugden, B., Sharp, P.A., Sambrook, J.: Proc. Natl. Acad. Sci. US 71, 4183 (1974)
7. Martin, M., Khoury, G.: Curr. Top. Microbiol. 73, 35 (1976)
8. Sambrook, J., Westphal, H., Srinivasan, P.R., Dulbecco, R.: Proc. Natl. Acad. Sci. US 60, 1288 (1968)
9. Shani, M., Rabinowitz, Z., Sachs, L.: J. Virol. 10, 456 (1972)
10. Sambrook, J., Sharp, P.A., Keller, W.: J. Mol. Biol. 70, 57 (1972)
11. Ozanne, B., Sharp, P.A., Sambrook, J.: J. Virol. 12, 90 (1973)
12. Khoury, G., Martin, M.A., Lee, T.N.H., Nathans, D.: Virology 63, 263 (1975)
13. Flint, S.J., Gallimore, P.H., Sharp, P.A.: J. Mol. Biol. 96, 47 (1975)
14. Tooze, J. (ed.): The Molecular Biology of Tumour Viruses. New York: Cold Spring Harbor Lab. 1973, pp. 375-381
15. Philipson, L., Lindberg, A.: Comprehensive Virology. Fraenkel-Conrat, H., Wagner, R.R. (eds.). New York; Plenum Press; 1974, pp. 143-207
16. Botchan, M.R., McKenna, G.: Cold Spring Harbor Symp. Quant. Biol. 38, 391 (1973)
17. Ketner, G., Kelly, T.J.: Proc. Natl. Acad. Sci. US 73, 1102 (1976)
18. Botchan, M., Topp, W., Sambrook, J.: Cell 9, 269 (1976)
19. Croce, C.M., Huebner, K., Girardi, A.J., Koprowski, H.: Cold Spring Harbor Symp. Quant. Biol. 39, 335 (1974)
20. Duff, R., Rapp, F.: J. Virol. 8, 469 (1971)
21. Rapp, F., Li, J.H., Jerkofsky, M.: Virology 55, 339 (1973)
22. Darai, F., Munk, K.: Nature (New Biol.) 241, 268 (1973)
23. Geder, L., Vaczi, L., Boldogh, I.: Acad. Sci. Hung. 20, 119 (1973)
24. Kutinova, L., Vonka, V., Broucek, J.: J. Natl. Cancer Inst. 50, 759 (1973)
25. Duff, R., Kreider, J.W., Levy, B.M., Katz, M., Rapp, F.: J. Natl. Cancer Inst. 53, 1159 (1974)
26. MacNab, J.C.M.: J. Gen. Virol. 24, 143 (1974)
27. Takahashi, M., Yamanishi, K.: Virology 61, 306 (1974)
28. Kimura, S., Flannery, V.L., Levy, B., Schaffer, P.A.: Intern. J. Cancer 15, 786 (1975)
29. Boyd, A.L., Orme, T.W.: Intern. J. Cancer 16, 526 (1975)
30. Kucera, L.S., Gusdon, J.P.: J. Gen. Virol. 30, 257 (1976)
31. Munyon, W., Kraiselburd, E., Davis, D., Mann, J.: J. Virol. 7, 813 (1971)
32. Davidson, R., Adelstein, S., Oxman, M.: Proc. Natl. Acad. Sci. US 70, 1912 (1973)
33. Duff, R., Rapp, F.: J. Virol. 15, 490 (1975)
34. Shin, S.I., Freedman, V.H., Risser, R., Pollack, R.: Proc. Natl. Acad. Sci. US 72, 4435 (1975)
35. Rapp, F., Duff, R.: Cancer Res. 33, 1527 (1975)
36. MacNab, J.C.M., Timbury, M.C.: Nature (London) 261, 233 (1976)
37. Kimura, S., Esparza, J., Benyesh-Melnick, M., Schaffer, P.A.: Intervirology 3, 162 (1974)
38. Lin, S.S., Munyon, W.: J. Virol. 14, 1199 (1974)
39. Leiden, J., Buttyan, R., Spear, P.G.: J. Virol. 20, 413 (1976)
40. Reed, C.L., Cohen, G.H., Rapp, F.: J. Virol. 15, 668 (1975)
41. Marquez, E.G., Rapp, F.: Intervirology 6, 64 (1975)
42. Flannery, V.L., Courtney, R.J., Schaffer, P.A.: J. Virol. 21, 284 (1977)
43. Honess, R.W., Roizman, B.: J. Virol. 12, 1347 (1973)
44. Collard, W., Thornton, H., Green, M.: Nature (New Biol.) 243, 264 (1973)
45. Copple, C.D., McDougall, J.K.: Intern. J. Cancer 17, 501 (1976)
46. Kraiselburd, E., Gage, L.P., Weissbach, A.: J. Mol. Biol. 97, 533 (1975)
47. Davis, D.B., Kingsbury, D.T.: J. Virol. 17, 788 (1976)
48. Minson, A.C., Thouless, M.E., Eglin, R.P., Darby, G.: Intern. J. Cancer 17, 493 (1976)
49. Frenkel, N., Locker, H., Cox, B., Roizman, B., Rapp, F.: J. Virol. 18, 885 (1976)
50. Martin, R.G., Chou, J.Y.: J. Virol. 15, 599 (1975)
51. Tegtmeyer, P.: J. Virol. 15, 613 (1975)
52. Brugge, J.S., Butel, J.S.: J. Virol. 15, 619 (1975)
53. Osborn, M., Weber, K.: J. Virol. 15, 636 (1975)
54. Kimura, G., Itagaki, A.: Proc. Natl. Acad. Sci. US 72, 673 (1975)
55. Roizman, B., Frenkel, N.: In: Proc. Anglo-Am. Conf. Sexually Transmitted Diseases. Royal Soc. Med., London. June 23-25, 1976.

A New Transforming Papovavirus Isolated from Primate Cells

G. Sauer, W. Waldeck, and K. Bosslet

The family of papovaviruses comprises small icosahedral DNA tumor viruses that contain circular double-stranded DNA and whose coat consists of capsomers. Furthermore, they are ether-resistant and capable of transforming cells in vitro. Papilloma, polyoma, and simian virus 40 (SV40) are the most prominent representatives of this group, together with two more recently recognized virus species called BK and JC, which were isolated from human patients (Gardner et al., 1971; Padgett et al., 1971).

We have isolated a further papovavirus designated HD from a permanent line of primate cells (Waldeck and Sauer, 1977). The VERO cell line, which harbors the virus, was originally derived from kidney cells of the African green monkey *Cercopithecus aethiops*. While HD virus can be recovered from the VERO cells that are being maintained in our laboratory, we have been unable to demonstrate the presence of the virus in VERO cells obtained from other sources. Hence, at present the origin of HD virus remains obscure.

HD virus grows without any apparent cytopathogenic effect both in our VERO cell line and, although poorly, in RITA cells (another line of *Cercopithecus aethiops* origin). We have not found, as yet, other host-cell systems that would support the vegetative growth of the virus. Since no plaque assay is available, the synthesis of radioactively labeled HD superhelical DNA is being used as a measure of infectivity.

Replication of HD DNA is most pronounced in growing VERO cells. Stationary cell cultures, in contrast, reveal smaller amounts of HD DNA (Fig. 1). For preparation of infectious virus, VERO cells were frozen and thawed, the cellular debris was pelleted, and the virus-containing supernatant was used, after treatment with DNase and with ether (to inactivate putative RNA tumor viruses), for infection.

When VERO cells from other laboratories that were proved to be virus-free were infected with lysates from our VERO cells, approximately 1% of the cells grew to form colonies in soft agar (Fig. 2) and between 0,7 - 1% of the infected cells were capable of forming colonies in medium containing only 2% serum. In contrast, 0,04% of the uninfected control cells formed colonies under the same conditions. Both properties, namely, growth in soft agar (Macpherson and Montagnier, 1964) and reduced serum requirement (Clarke et al., 1970; Dulbecco, 1970) show that HD virus is capable of transforming VERO cells.

The HD genome consists of two size classes of superhelical DNA with molecular weight of either 3.45×10^6 and 3.25×10^6 daltons. While SV40, BK, and polyoma papova viruses share some base sequences homology with each other (Khoury et al., 1975; Osborn et al., 1976; Ferguson and Davis, 1976), HD DNA is devoid of such homologies, as shown in DNA-DNA hybridization experiments (only 1.2% homology with SV40).

The HD genome is also distinguished from the other papova genomges by its restriction enzyme cleavage pattern. A physical map of the HD

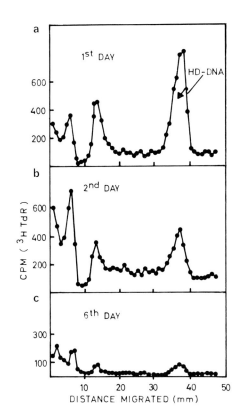

a 1st DAY HD-DNA

b 2nd DAY

c 6th DAY

CPM (^{3}H TdR)

DISTANCE MIGRATED (mm)

Fig. 1a - c. Dependence of HD viral DNA replication on cellular growth. VERO cells in parallel cultures were labeled with ^{3}H-thymidine for 12 h (a) when they were actively growing one day after seeding; (b) before reaching confluency at day 2 and, (c) when they were confluent at day 6. After labeling, the viral DNA was selectively isolated and electrophoresed in 1.4% agarose gels as described (Waldeck et al., 1976). After electrophoresis the gels were sliced and the radioactivity was determined. The peaks migrating at fraction 37 represent superhelical HD DNA

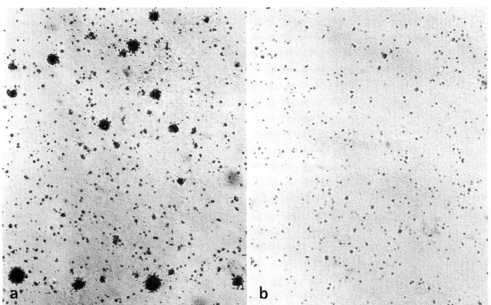

Fig. 2a and b. Soft agar assays of (a) HD-infected and (b) uninfected VERO cells

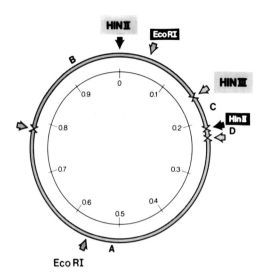

Fig. 3. Physical map of the HD virus genome

genome is shown in Figure 3. It can be seen that restriction endonu-
clease EcoRI cuts the DNA twice (unlike SV40, which is cut only once).
Five fragments are generated by the restriction endonucleases HindII
+ HindIII.

In the smaller HD DNA fragments C and D are missing. The cleavage site
of HindII which cuts the smaller HD DNA only once, has been chosen as
the starting position of the map. Based on the following taxonomic
criteria, HD virus is a new member of the papovavirus group: its
genome consists of circular closed double-stranded DNA, its morphology
as revealed by electron micrographs is very similar to the size and
morphology of SV40, it is ether-resistant, and it is capable of transforming VERO cells. Since its DNA is not homologous to the DNA of
SV40, BK, and polyoma, we have isolated a virus whose DNA can serve
as a new probe in the search for homologies in tumors of various
origins.

References

Clarke, G.D., Stoker, M.G.P., Ludlow, A., Thornton, M.: Nature (London) 227, 798
    (1970)
Dulbecco, R.: Nature (London) 227, 802 (1970)
Ferguson, J., Davis, R.W.: J. Mol. Biol. 94, 496 (1976)
Gardner, S.D., Field. A.M., Coleman, D.V., Hulme, B.: Lancet, i: 1253 (1971)
Khoury, G., Howley, P.M., Garon, C., Mullarkey, M.F., Takemoto, K.K., Martin, M.A.:
    Proc. Natl. Acad. Sci. USA 72, 2563 (1975)
Macpherson, I., Montagnier, L.: Virology 23, 291 (1964)
Osborn, J.E., Robertson, S.M., Padgett, B.L., Walter, D.L., Weisblum, B.: J. Virol.
    19, 675 (1976)
Padgett, P.L., Walter, D.L., Zu Rhein, G.M., Eckroade, R.J., Dessel, B.H.: Lancet, i:
    1257 (1971)
Waldeck, W., Chowdhury, K., Gruss, P., Sauer, G.: Biochim. Biophys. Acta 425, 157
    (1976)
Waldeck, W., Sauer, G.: Nature, 269, 171 (1977)

Selectivity in the Integration of Viral DNA in Cells Infected and Transformed by Adenovirus

E. Fanning, K. Baczko, D. Sutter, and W. Doerfler

Introduction

In a series of studies we have demonstrated that the DNA of human
adenovirus types 2 and 12 can be covalently linked to cellular DNA
(for reviews see Doerfler, 1977; Doerfler et al., 1977). Recombination
between viral and cellular DNA leading to the insertion of viral DNA
sequences into the host genome appears to be a general phenomenon occurring in abortively infected (Doerfler, 1968; 1970), in productively
infected (Burger and Doerfler, 1974; Doerfler et al., 1974; Schick et
al., 1976), and in transformed cells (Bellett, 1975; Green et al.,
1976; Groneberg et al., 1977). Viral DNA sequences are integrated in
the form of fragments. There is evidence that viral DNA replication
is not required for insertion (Doerfler, 1970; Fanning et al., 1978),
whereas the integrated form of adenovirus DNA cannot be detected when
cellular DNA is completely inhibited. The mechanism by which viral and
cellular sequences are linked is not at all understood. It is likely
that cellular replication and recombination mechanisms play an important
role in the insertion.

In investigations on the integration of adenovirus DNA, two approaches
have proved particularly useful:

1.   Studies of the very early steps involved in the interaction between adenovirus type 2 (Ad2) and permissive human (KB) cells (Burger
and Doerfler, 1974; Schick et al., 1976) or between adenovirus type
12 (Ad12) and nonpermissive hamster (BHK21) cells (Doerfler, 1968;
1970) have been performed. These systems may help reveal details of
the mechanisms by which insertion proceeds.

2.   In established lines of adenovirus-transformed hamster cells the
Ad12 genome persists in specific patterns (Fanning and Doerfler, 1976;
Green et al., 1976) in an integrated state (Groneberg et al., 1977).
Early Ad12-specific functions continue to be expressed in these cells
(Ortin et al., 1976). The transformed cell lines represent, as it
were, a terminal state in the interaction between the host and viral
genomes. In transformed cells viral genes have become part of the host
genome. The transformed cell system permits investigations on the
effects of the persisting viral genes on host functions.

The question of specificity in terms of which fragments of viral DNA
are integrated and what cellular sites become occupied, is of para-
mount importance. In this communication we demonstrate that in Ad2-
infected KB cells the right molecular end of Ad2 DNA is represented
at least ten to fifteen times more frequently in the integrated form
than the left terminus. It has been shown previously that in productively infected cells integrated viral DNA sequences can be detected
very early after infection (Schick et al., 1976; Fanning and Doerfler,
1977) and that probably parental viral DNA is linked to cellular DNA
(Groneberg et al., 1975). Furthermore, it will be shown that viral
DNA sequences can be excised from the high-molecular-weight DNA from
Ad2-infected KB cells.

In four lines of Ad12-transformed hamster cells, which have been shown to produce tumors in animals, the patterns of integrated viral DNA sequences in cellular DNA have been determined by the techniques developed by Southern (1975) and by Ketner and Kelly (1976). In each line viral DNA sequences have been detected in a limited number of sites, and these sites are not identical in the four Ad12-transformed hamster cell lines.

Experimental details will not be discussed in this communication, although the legends to figures will provide some technical information. The reader will be referred to reports published previously or in press.

Integration of Adenovirus DNA in Human Cells Productively Infected with Ad2

Size Class Analysis of Newly Synthesized DNA in Ad2-Infected KB Cells

As described earlier (Burger and Doerfler, 1974; Doerfler et al., 1974; Schick et al., 1976), in KB cells productively infected with Ad2, four size classes of newly synthesized viral DNA have been found by zone velocity sedimentation in alkaline sucrose density gradients (Fig. 1): Ad2 DNA in the >100 S, the 50 - 90 S, the 34 S (genome size DNA) and the <20 S regions. The >100 S, the 50 - 90 S and the <20 S size classes of newly synthesized DNA are also observed when the DNA from mock-infected KB cells is analyzed by the same method. Evidence has been presented that in uninfected cells, the ^{3}H label in the 40-100 S DNA can be chased into the >100 S DNA (Schick et al., 1976). The presence of Ad2 DNA sequences in the four size classes of DNA has been documented by DNA-DNA filter hybridization (Schick et al., 1976).

Quantitation of Total Intracellular Ad2 DNA

KB cells productively infected with Ad2 are a very efficient system for the replication of viral DNA. The total number of Ad2 DNA equivalents per cell at various times after infection was determined by reassociation kinetics. ^{3}H-labeled Ad2 DNA was reassociated with unlabeled DNA from a known number of uninfected cells or in the presence of the total intracellular DNA extracted from the same number of Ad2-infected KB cells extracted at various times after infection. Late after infection, up to 400,000-500,000 Ad2 DNA equivalents per cell are found.

Quantitation of Viral DNA in Four Size Classes

At various times after infection, the total intracellular DNA from unlabeled cells was fractionated by zone velocity sedimentation in alkaline sucrose density gradients (Schick et al., 1976). The DNA from individual size classes was pooled and subjected to reassociation analysis to determine the amount of Ad2 DNA using ^{3}H-labeled Ad2 DNA as probe. The number of Ad2 DNA equivalents found in the four size classes at 8, 12, 18, and 30 h after infection are listed in Table 1. Relative to the amount of Ad2 DNA of unit genome size, the number of viral genome equivalents in the high-molecular-weight (HMW) region (i.e., the >100 S and the 50 - 90 S size classes) is higher at early times after infection (p.i.), e.g., about 26% at 8 h p.i., than late after infection, e.g., 4 - 5% at 30 h p.i. This finding may indicate

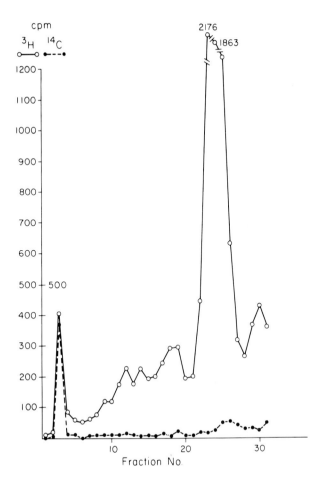

Fig. 1. DNA synthesis in Ad2-infected KB cells: Analysis by zonal sedimentation in alkaline sucrose density gradients. KB cells growing in monolayers were prelabeled by adding ^{14}C-thymidine (0.4 µCi/ml) to the medium and maintaining the cells under these conditions for 3 days prior to infection. Immediately prior to infection, the cells were washed several times with phosphate-buffered saline (PBS) to remove free ^{14}C-thymidine from the cells. From 12 - 18 h after infection, the newly synthesized DNA was labeled with ^{3}H-thymidine (30 µCi/ml medium). At the end of the labeling period, the cells were washed several times with PBS and were lysed for 18 h on top of an alkaline sucrose gradient. The gradient was centrifuged at 22,000 rpm for 380 min at 4°C in the SW27 rotor of the L2-65B ultracentrifuge. This figure was taken from Doerfler et al. (1974). Reprinted with permission of Cold Spring Harbor Laboratory

that a significant proportion of the parental viral DNA is converted to the HMW forms of Ad2 DNA.

Is the Entire Viral Genome Represented in the Four DNA Size Classes?

This question can be answered by using ^{3}H-labeled, specific restriction endonuclease fragments of Ad2 DNA in reassociation experiments with DNA from the four size classes isolated from Ad2-infected KB

Table 1. Quantitation of Ad2 DNA in four size classes at different times after infection[a]

Time of analysis, h after infection of KB cells with Ad2	Unlabeled DNA from Ad2-infected KB cells sedimenting at	Cell number in reaction mixture	Concentration of ^{3}H-labeled Ad2 probe DNA (OD_{260}/ml)	$\dfrac{C_0t_{1/2} \text{ control}}{C_0t_{1/2} \text{ experimental}}$	Ad2 DNA equivalents per cell
8 h p.i.	>100 S	not analyzed			
	50 – 90 S	2.6×10^5	2.33×10^{-4}	1.54	1200
	34 S	2.3×10^5	2.56×10^{-4}	2.22	3400
	< 20 S	2.9×10^5	2.75×10^{-4}	1.0	ND[b]
12 h p.i.	>100 S	2.9×10^5	2.61×10^{-4}	1.48	1100
	50 – 90 S	6.0×10^5	2.38×10^{-4}	2.35	1300
	34 S	4.7×10^5	2.39×10^{-4}	5.22	5300
	< 20 S	8.1×10^5	2.65×10^{-4}	1.58	500
18 h p.i.	>100 S	6.3×10^5	4.02×10^{-4}	1.0	ND
	50 – 90 S	1.3×10^5	9.01×10^{-4}	1.42	750
	34 S	6.3×10^5	6.79×10^{-4}	2.81	4800
	< 20 S	5.4×10^5	4.27×10^{-4}	2.34	2600
30 h p.i.	>100 S	8.2×10^5	3.96×10^{-4}	3.12	2500
	50 – 90 S	9.6×10^5	3.76×10^{-4}	2.13	1100
	34 S	6.0×10^5	3.73×10^{-4}	43.9	66000
	< 20 S	1.2×10^5	3.54×10^{-4}	22.3	16000

[a] DNA from Ad2-infected or mock-infected cells was isolated by alkali lysis, and DNA in the four size classes was isolated by zone velocity sedimentation in alkaline sucrose density gradients. The amount of Ad2 DNA in each size class was determined by reassociation kinetics using ^{3}H-labeled Ad2 DNA (spec. act. 2.02×10^7 cpm/OD_{260} unit) as probe
[b] ND: none detected
This table was taken from Fanning and Doerfler (1977). Reprinted with permission of Academic Press

cells. Each preparation of a particular size class was also analyzed for viral DNA sequences using ^{3}H-labeled Ad2 unit size DNA as the probe. The amount of each fragment of Ad2 DNA present was related to the amount of intact Ad2 DNA. Thus the relative amount of each of the fragments in that size class could be compared from one preparation to another. In this way the relative abundance of each of the fragments in each of the size classes has been determined. Ad2 DNA was cleaved with the EcoRI or BamHI restriction endonuclease. The positions of these restriction sites on the Ad2 DNA molecule are depicted schematically in Figure 2.

As expected, the EcoRI restriction endonuclease fragments A-F are represented in about equimolar amounts in the 34 S unit genome size Ad2 DNA peak. The EcoRI A fragment is possibly slightly less abundant than the other parts of the genome (Fig. 3). The data obtained from reassociation measurements for each restriction endonuclease fragment have been normalized relative to the total amount of viral DNA present and have been corrected for the fractional length of each fragment.

The DNA isolated from the 50 – 90 S region of the alkaline sucrose density gradients has been subjected to the same analysis using frag-

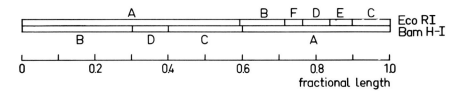

Fig. 2. Map of the EcoRI and BamHI fragments of Ad2 DNA. The map positions have been determined by Pettersson et al.(1973), for the EcoRI fragments and by R. Roberts (personal communication) for the BamHI fragments. The scale on the bottom indicates fractional length units

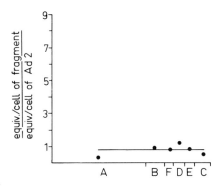

Fig. 3. Relative abundance of each of the EcoRI fragments of Ad2 DNA in the 34 S DNA peak from Ad2-infected KB cells. Unlabeled 34 S (unit genome size) DNA was prepared 18 - 20 h p.i. by zone velocity sedimentation on alkaline sucrose density gradients as described (Fanning and Doerfler, 1977). This DNA was analyzed for viral DNA sequences by reassociation kinetics using each of the EcoRI restriction fragments of ^{3}H-labeled Ad2 DNA as probe (spec. act. 3.05×10^{7} cpm/OD_{260} unit). Each preparation was also analyzed using ^{3}H-labeled Ad2 virion DNA (spec. act. 2.02×10^{7} cpm/OD_{260} unit) as probe. The number of fragment equivalents per cell found in 34 S DNA was divided by the number of equivalents per cell found in 34 S DNA when the whole Ad2 genome was used as probe. The relative abundance of the fragments is shown on the ordinate and the relative size and location of the EcoRI fragments on the Ad2 DNA are shown on the abscissa. Details of the composition of the reassociation mixtures are given in Fanning and Doerfler (1977), from which this Figure was taken. Reprinted with permission of Academic Press

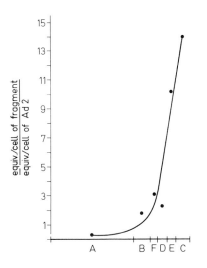

Fig. 4. Relative abundance of each of the EcoRI fragments of Ad2 DNA in the 50 - 90 S DNA from Ad2-infected KB cells. Experimental conditions were as described in the legend to Figure 3, except that unlabeled 50 - 90 S DNA was analyzed. This Figure was taken from Fanning and Doerfler (1977). Reprinted with permission of Academic Press

ments of Ad2 DNA that were generated by the restriction endonuclease EcoRI (Fig. 4). The data demonstrate that in the 50 - 90 S DNA the right end of the Ad2 DNA molecule occurs at least ten to fifteen times more abundantly than the left terminus. In fact, there is a striking gradient in the frequency of occurrence in the 50 - 90 S DNA of Ad2 DNA fragments from the left molecular terminus to the right molecular end. Similar results are obtained when the BamHI fragments are used in this type of analysis (data not shown).

The finding of overrepresentation of the right part of the Ad2 DNA molecule conclusively rules out the possibility that Ad2 DNA sequences in the HMW forms of newly synthesized DNA in Ad2-infected KB cells could be due to artifactual drag, trapping, or non-specific association.

In the <20 S size class of DNA the right half of the Ad2 DNA molecule is overrepresented by a factor of 2 - 4, depending on the location on the viral DNA map. Lastly, the >100 S DNA isolated from Ad2-infected KB cells was also investigated by the technique described. In this size class, the left half of the molecule appears to be more abundant (data not shown).

Considering the relative total amounts of Ad2 DNA sequences in the >100 S, 50 - 90 S, and <20 S DNA size classes (Table 1) and the frequency gradients for individual fragments of Ad2 DNA observed in each size class, there remains a considerable net excess of the right molecular end of the Ad2 DNA molecule in productively infected KB cells.

Excision of Viral DNA Sequences from High-molecular-weight Sequences

Experimental Design

If indeed the HMW forms of viral DNA from Ad2-infected KB cells represent viral DNA sequences integrated into cellular DNA, it should be possible to excise these sequences from the HMW cellular DNA with restriction endonucleases. In this case, analysis of the fragments separated by electrophoresis on gels should reveal viral sequences in size classes different from those characteristic of virion DNA. The approach of excising viral sequences has been successfully applied to the characterization of integrated viral sequences in SV40 (Botchan and McKenna, 1973; Botchan et al., 1976) and adenovirus type 12-transformed cells (Groneberg et al., 1977). This experimental plan also permits one to test for the occurrence of simple oligomeric forms of viral DNA, since according to this model adducts of the terminal fragments of viral DNA should be found.

a) Restriction Endonuclease EcoRI: Cells productively infected with Ad2 were labeled with ^{3}H-thymidine from 6 - 12 h p.i.; the total intranuclear DNA was extracted, the HMW DNA was isolated by zone velocity sedimentation on neutral sucrose density gradients (Fig. 5a) and subsequently cleaved with the restriction endonuclease EcoRI. ^{14}C-labeled Ad2 virion DNA was added as an internal marker prior to cleavage to document that all the DNA had been completely fragmented. The fragments were then separated on gels, and the DNA from individual gel slices was analyzed by DNA-DNA filter hybridization using intact Ad2 DNA on the filters. The ^{14}C-labeled Ad2 virion DNA served as an internal marker to monitor the efficiency of the hybridization reaction. The results of this experiment indicate that upon cleavage with the EcoRI restriction endonuclease, the bulk of the ^{3}H-labeled cellular DNA is distributed in size classes between the EcoRI A (mol. w.

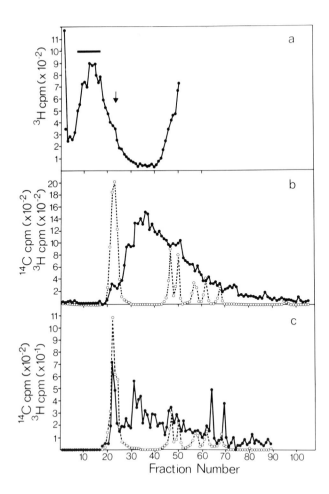

Fig. 5a – c. Analysis after gel electrophoresis of the EcoRI restriction enzyme fragments of the cellular DNA peak synthesized 6 – 12 h p.i. in Ad2-infected KB cells.
(a) Distribution of the ^{3}H-labeled DNA synthesized 6 – 12 h p.i. in a neutral sucrose
gradient. Conditions of centrifugation: SW41 rotor, 35,000 rpm, 4°, 360 min.
(b) Fractions 7 – 17, indicated by the horizontal bar (a), were pooled, dialyzed, and
concentrated. Subsequently, 2 – 5 μg of ^{14}C-labeled Ad2 DNA was added and the mixture
was incubated with the EcoRI restriction endonuclease. The fragmented DNA was separated by electrophoresis on polyacrylamide-agarose gels. The labeled DNA from individual gel slices was eluted, and 0.1-ml portions were dried on nitrocellulose
filters and counted in toluene-based scintillator. ^{3}H-labeled DNA (cpm) (●); ^{14}C-
labeled Ad2 marker DNA (cpm) (o) representing the six specific EcoRI fragments.
(c) The DNA from each gel slice was hybridized to Ad2 DNA on filters. The ^{3}H- and
^{14}C-activity curves (symbols as in b) represent the positions of Ad2-specific sequences. This Figure was taken from Schick et al. (1976). Reprinted with permission
of the National Academy of Sciences, USA

13.6 × 10^6) and the EcoRI F (mol. w. 1.1 × 10^6) fragments of Ad2 DNA
(Fig. 5b). Analysis of the DNA in each gel slice by DNA-DNA hybridization to intact Ad2 DNA on filters reveals that viral-specific sequences
are located in the positions of most, if not all the EcoRI Ad2 marker
DNA fragments and also in the regions between EcoRI fragments A and B,

and C and D (Fig. 5c). Thus a considerable amount of the Ad2-specific DNA from the HMW peak of cellular DNA is found upon cleavage with the restriction endonuclease EcoRI in size classes that do not coincide with the fragments of the virion marker DNA. Moreover, it is striking that viral specific sequences cannot be detected in DNA of size exceeding that of the EcoRI A fragment (Fig. 5c). This latter finding has been confirmed by the results of zone velocity sedimentation experiments in neutral sucrose gradients.

b) Restriction Endonuclease BamHI: A similar set of experiments was carried out using the restriction endonuclease BamHI. When HMW cellular DNA from Ad2-infected KB cells was mixed with ^{14}C-labeled Ad2 marker DNA, cleaved with the BamHI restriction endonuclease, and subsequently analyzed by electrophoresis on polyacrylamide-agarose gels, a characteristic pattern was obtained (Fig. 6a). In contrast to the results observed after cleavage with EcoRI endonuclease, cellular DNA was found of a molecular weight higher than that of the BamHI A fragment. These findings are not contradictory, since 1) different restriction endonucleases would be expected to generate different cleavage patterns, and 2) the molecular weight of the EcoRI A fragment (13.6×10^6 daltons) is considerably higher than that of the BamHI A fragment (approximately 8.8×10^6 daltons). Again the DNA from each of the gel fractions was analyzed by hybridization to Ad2 DNA on filters. Viral DNA sequences are found in size classes corresponding to most of the BamHI fragments of Ad2 DNA. The DNA of a molecular weight higher than that of the BamHI A fragment showed considerable homology to Ad2 DNA (Fig. 6b) and to each of the BamHI fragments of Ad2 DNA (data not shown). Thus the restriction endonuclease BamHI, similar to EcoRI, produces DNA fragments containing viral sequences in size classes different from those of the virion DNA fragments.

These findings demonstrate that in HMW cellular DNA from Ad2-infected KB cells viral sequences are linked to DNA that is different from virion DNA, i.e., to cellular DNA. Also, when the restriction endonuclease SalI was used, fragments of DNA containing viral sequences were generated that were considerably larger than any of the SalI fragments of Ad2 marker DNA.

Conclusions

In KB cells productively infected with adenovirus type 2, viral DNA sequences can be detected in high-molecular-weight DNA size classes, starting 8 h, and probably 4 h postinfection. The amount of viral DNA in these size classes remains rather constant throughout infection. In the 34 S virion size DNA, all segments of the Ad2 genome are about equally represented, whereas the 50 - 90 S DNA contains the right terminus of the Ad2 genome at least ten to fifteen times more abundantly than the left one. In the <20 S DNA the right terminus is also overrepresented, though less strikingly, and in the >100 S DNA a preponderance of the left molecular end is observed. These data are best explained by a model in which fragments of Ad2 DNA are linked to cellular DNA. The right terminus of the Ad2 genome is preferentially involved in integration. The integrated viral DNA sequences can be excised from the HMW DNA in Ad2-infected KB cells with restriction endonucleases. Using three such enzymes, EcoRI, BamHI, and SalI, fragments are generated that consist of viral sequences linked to nonviral, most likely cellular DNA. The results of experiments in which the EcoRI endonuclease has been used rule out the possibility that HMW viral DNA consists of simple oligomeric forms (concatenates) of Ad2 DNA.

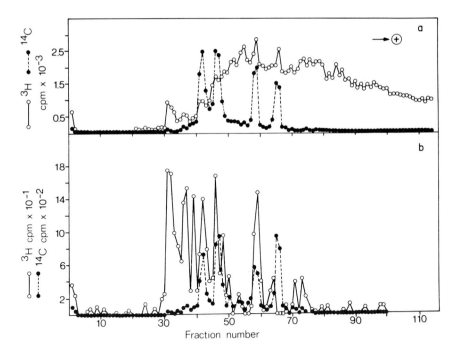

Fig. 6a and b. Analysis of the high-molecular-weight DNA from Ad2-infected KB cells after electrophoresis on polyacrylamide-agarose gels after cleavage with the BamHI restriction endonuclease.
Experimental conditions were similar to those described in the legend to Figure 5, except that the Ad2-infected KB cells were labeled with ^{3}H-thymidine 6-10 h p.i. and that the total nuclear DNA was cleaved with the BamHI restriction endonuclease. (a) Separation of the ^{3}H-labeled cellular (o) and ^{14}C-labeled viral (●) DNA fragments by electrophoresis on a polyacrylamide-agarose gel. (b) The DNA from each gel slice was analyzed by hybridization to intact Ad2 DNA. This Figure was taken from Baczko, Neumann, and Doerfler, manuscript submitted for publication (1978)

The evidence presented here together with the results obtained earlier (Burger and Doerfler, 1974; Schick et al., 1976) supports the notion that fragments of viral DNA become linked to cellular DNA early in permissive infection. It is possible that the amount of integrated viral sequences is amplified by the replication of cellular DNA. The function of the integrated viral sequences in virus replication and the consequences of the numerous integration events for the host cell remain to be determined.

Patterns of Integration of Viral DNA Sequences in Adenovirus Type 12-transformed Hamster Cells

Four different lines of adenovirus type 12-transformed hamster cells have been investigated. The properties of the lines T637, HA12/7, A2497-2, and A2497-3 are summarized in Table 2. It was shown previously that the entire viral genome is present in multiple copies in each of these cell lines and that different segments of the Ad12 genome are represented at different frequencies (Fanning and Doerfler, 1976).

Table 2. Properties of Ad12-transformed hamster cell lines

Cell line	Cells used for transformation	Multiplicity of Ad12 used in transformation	T-Antigen	Oncogenicity in animals[a] 10^5 10^6 cells/animal		Reference
T637	BHK21	350	+	5/10	7/10	Strohl et al., (1970)
HA12/7	Primary Syrian hamster	10	+	1/10	4/10	zur Hausen (1973)
A2497-2	LSH inbred hamster embryo	5	+	7/10	10/10	A.M. Lewis, Jr.
A2497-3	" "	5	+	6/10	5/10	personal communication

[a] The number of cells indicated was injected subcutaneously into Syrian hamsters (50 - 60 g). The first tumors were observed 3 - 4 weeks after the injection. The figures indicate the number of animals in which tumors formed/number of animals injected. Histologically the tumors were undifferentiated round cell sarcomas that showed noninfiltrative growth behavior.
With modifications this Table was taken from Doerfler et al. (1977)

Thus at least part of the viral DNA must persist in the form of fragments. It has been shown by restriction enzyme analysis and by sequential hybridization experiments that the viral DNA persists in all four cell lines in an integrated form (Groneberg et al., 1977).

The pattern in which Ad12 DNA is integrated at certain sites of the hamster cell genome has been investigated by the technique of Ketner and Kelly (1976) using nick-translated (Rigby et al., 1977) Ad12 DNA. The details of the procedures used have been outlined in Figure 7 and in the legend to Figure 8. The DNA of the Ad12-transformed lines T637, HA12/7, A2497-2, and A2497-3 was cleaved with the EcoRI (Fig. 8a) or the BamHI (Fig. 8b) restriction endonuclease. DNA fragments were separated by electrophoresis on agarose gels, transferred to nitrocellulose filters, and Ad12 DNA sequences in the cellular DNA fragments were detected by hybridization with ^{32}P-labeled Ad12 DNA and autoradiography. The results (Fig. 8a, b) are:

1. Upon digestion with the EcoRI (Fig. 8a) or BamHI (Fig. 8b) restriction endonuclease, a distinct pattern of a limited number of integration sites is observed for each cell line investigated. The DNA from BHK21 cells or the B3 line of BHK21 cells does not contain viral DNA sequences.

2. The patterns are different for each cell line studied.

3. Judging from the size of the Ad12-specific sequences, the terminal fragments A and C, upon digestion with the EcoRI endonuclease (Fig. 8a), and A und E, upon digestion with the BamHI endonuclease (Fig. 8b), do not seem to be present in the molecular weight regions of the gel in which these fragments would be found had unit length virion Ad12 DNA been cleaved (cf. Fig. 8c).

4. Upon digestion with the BamHI endonuclease, fragments of cellular DNA containing viral sequences are generated with DNA from all four lines, some of which have a higher molecular weight than the BamHI A fragment (mol. wt. 4.9×10^6). For comparison, the EcoRI and BamHI maps of Ad12 DNA are shown in Figure 8c.

BLOTTING TECHNIQUE

1) CLEAVE DNA FROM TRANSFORMED CELLS WITH RESTRICTION ENDONUCLEASES

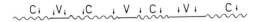

2) SEPARATE FRAGMENTS ON 1% AGAROSE SLAB GELS.

3) DENATURE DNA IN GEL BY 1N KOH, NEUTRALIZE AND TRANSFER DNA
BY BLOTTING TO NITROCELLULOSE FILTER (A).

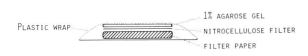

PLASTIC WRAP ···········
···· 1% AGAROSE GEL
···· NITROCELLULOSE FILTER
···· FILTER PAPER

4) INCUBATE FILTER WITH DENHARDT MIXTURE FOR 15-20H, HYBRIDIZE TO
^{32}P-LABELED VIRUS DNA LABELED BY NICK-TRANSLATION (B).

5) WASH FILTERS EXTENSIVELY, EXPOSE TO X-RAY FILM, DEVELOP.

Fig. 7. Blotting technique to localize specific genes within cellular DNA. In this
set of experiments, fragments of integrated Ad12 genes were characterized as to
their location in specific fragments of hamster cell DNA. In 1) C refers to cellular,
V to integrated viral DNA sequences. In 2) DNA fragments are separated by electrophoresis on 1% agarose slab gels. References given in the scheme are (A) to Southern
(1975), and (B) to Rigby et al.(1977). Details of this procedure have been described
by Ketner and Kelly (1976) and by Botchan et al. (1976)

Conclusions

Analysis of the integration pattern of Ad12 DNA in transformed hamster
cell lines is at an early stage. The data obtained so far with two
restriction endonucleases indicate, however, that each of the four
lines investigated exhibits a different pattern of integrated Ad12
sequences. This conclusion is strengthened by the results obtained
with an additional four endonucleases (Sutter, Westphal, and Doerfler,
manuscript submitted). The sequences located at the termini of the
Ad12 DNA molecule appear to be transposed to DNA fragments in different
size classes, i.e., are linked to cellular DNA. In general, there is
not perfect coincidence in size between the known virion DNA fragments
and the fragments containing viral DNA in each of the cell lines.

Considering the limited amount of data available on the integration
of Ad12 DNA in transformed hamster cells, we should like to reserve
judgement on the question of specific integration sites. In terms of
the high molecular weight of mammalian cell DNA, the patterns of integration reveal a certain degree of selectivity, but there are distinct

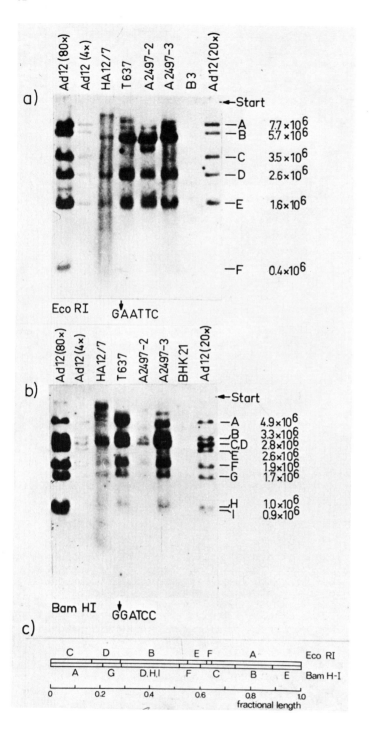

a)

Ad 12 (80×)   Ad12 (4×)   HA12/7   T 637   A 2497-2   A 2497-3   B 3   Ad 12 (20×)

←Start

—A   7.7 ×10^6
—B   5.7 ×10^6
—C   3.5 ×10^6
—D   2.6 ×10^6
—E   1.6 ×10^6

—F   0.4 ×10^6

Eco RI   G↓AATTC

b)

Ad12 (80×)   Ad12 (4×)   HA12/7   T 637   A 2497-2   A 2497-3   BHK 21   Ad 12 (20×)

←Start

—A   4.9 ×10^6
—B   3.3 ×10^6
—C,D 2.8 ×10^6
—E   2.6 ×10^6
—F   1.9 ×10^6
—G   1.7 ×10^6

—H   1.0 ×10^6
—I   0.9 ×10^6

Bam HI   G↓GATCC

c)

| C | D | B | E | F | A | | Eco RI |
| A | G | D,H,I | F | C | B | E | Bam H-I |

0   0.2   0.4   0.6   0.8   1.0
fractional length

Fig. 8a - c. Localization at specific sites of Ad12 DNA sequences in the DNA of Ad12-transformed hamster cell lines T637, HA12/7, A2497-2, and A2497-3. DNA from T637, HA12/7, A2497-2, or A2497-3 cells, and Ad12 DNA as an internal control, in amounts corresponding to 4,20 or 80 genome equivalents per cell, was incubated with the EcoRI (a) or BamHI (b) restriction endonuclease, and the DNA fragments were separated by electrophoresis on 1% agarose slab gels. As controls the DNAs from BHK21 cells (b) or from the B3 line of BHK21 cells (a) were treated in the same way. As indicated in the scheme in Figure 7, DNA fragments were then denatured in the gel and transferred in situ to nitrocellulose filters. Ad12-specific sequences were detected on the filters by hybridization with Ad12 DNA labeled with ^{32}P-dTTP by nick-translation, followed by autoradiography using Kodak XR-5 X-ray film. Details of the procedure will be found in Sutter, Westphal, and Doerfler (manuscript submitted for publication).

The amount of Ad12 DNA used in the reconstitution experiments corresponds to 4,20 or 80 genome equivalents per cell. Ten μg of cellular DNA has been used in each slot.

The molecular weights of the EcoRI (A-F) and BamHI (A-I) fragments of Ad12 DNA have been indicated (Mulder et al., 1974; Ortin et al., 1976), and the location of the fragments on the Ad12 genome is shown in (c). The sequences recognized by the EcoRI and BamHI endonucleases are also shown. The arrows in (a) and (b) indicate the start line for electrophoresis

differences among the four lines studied. However, there is no information on the arrangement of cellular genes in the DNA of the different hamster cell lines. It is conceivable that in different transformed hamster cell lines equivalent genes are localized at different sites, and thus viral genes appearing in different size classes on the gel could still be adjacent to similar cellular genes. Hence it is impossible to evaluate the specificity of viral integration sites on the basis of the size of viral specific sequences alone. Additional parameters are urgently needed in this analysis.

Acknowledgements. We should like to thank Marianne Stupp and Ute Winterhoff for technical assistance in part of this work. This research was supported by the Deutsche Forschungsgemeinschaft (SFB74) and by a grant from the Ministry of Science and Research of the State of North-Rhine-Westfalia (IIB8-6440).

References

Baczko, K., Neumann, R., Doerfler, W.: Virology submitted for publication
Bellett, A.J.D.: Virology 65, 427 (1975)
Botchan, M., McKenna, G.: Cold Spring Harbor Symp. Quant. Biol. 38, 391 (1973)
Botchan, M., Topp, W., Sambrook, J.: Cell 9, 269 (1976)
Burger, H., Doerfler, W.: J. Virol. 13, 975 (1974)
Doerfler, W.: Proc. Natl. Acad. Sci. USA 60, 636 (1968)
Doerfler, W.: J. Virol. 6, 652 (1970)
Doerfler, W.: In: Comprehensive Virology. Fraenkel-Conrat, H., Wagner, R.R. (eds.), Vol. 10, 279 (1977)
Doerfler, W., Baczko, K., Burger, H., Fanning, E., Groneberg, J., Ortin, J., Scheidtmann, K.H., Schick, J., Soboll, H. Bull. Inst. Pasteur, Paris 75, 141 (1977)
Doerfler, W., Burger, H., Ortin, J., Fanning, E., Brown, D.T., Westphal, M., Winterhoff, U., Weiser, B., Schick, J.: Cold Spring Harbor Symp. Quant. Biol. 39, 505 (1974)

Fanning, E., Doerfler, W.: J. Virol. 20, 373 (1976)
Fanning, E., Doerfler, W.: Virology 81, 433 (1977)
Fanning, E., Schick, J., Doerfler, W.: Virology, submitted for publication
Green, M.R., Chinnadurai, G., Mackey, J.K., Green, M.: Cell 7, 419 (1976)
Groneberg, J., Brown, D.T., Doerfler, W.: Virology 64, 115 (1975)
Groneberg, J., Chardonnet, Y., Doerfler, W.: Cell 10, 101 (1977)
Ketner, G., Kelly, T.J., Jr.: Proc. Natl. Acad. Sci. USA 73, 1102 (1976)
Mulder, C., Sharp, P.A., Delius, H., Pettersson, U.: J. Virol. 14, 68 (1974)
Ortin, J., Scheidtmann, K.H., Greenberg, R., Westphal, M., Doerfler, W.: J. Virol. 20, 355 (1976)
Pettersson, U., Mulder, C., Delius, H., Sharp, P.A.: Proc. Natl. Acad. Sci. USA 70, 200 (1973)
Rigby, P.W.J., Dieckmann, M., Rhodes, C., Berg, P.: J. Mol. Biol. 113, 237 (1977)
Schick, J., Baczko, K., Fanning, E., Groneberg, J., Burger, H., Doerfler, W.: Proc. Natl. Acad. Sci. USA 73, 1043 (1976)
Southern, E.M.: J. Mol. Biol. 98, 503 (1975)
Strohl, W.A., Rouse, H., Teets, K., Schlesinger, R.W.: Arch. ges. Virusforsch. 31, 93 (1970)
zur Hausen, H.: Progr. exp. Tumor Res. 18, 240 (1973)

Synthesis and Integration of Virus-specific DNA in Cells Infected by RNA Tumor Viruses

H. E. Varmus, P. R. Shank, H.-J. Kung, S. Hughes, R. Guntaka, S. Heasley, and J. M. Bishop

Introduction

After infection of either permissive or nonpermissive cells by RNA
tumor viruses, viral genetic information appears to persist indefi-
nitely in the form of virus-specific, double-stranded DNA integrated
covalently into the host chromosome (Varmus et al., 1975; Weinberg,
1977). In this report, we summarize our progress in deciphering the
sequence of molecular events that culminates in the integration of
viral DNA into the DNA of avian cells infected by avian sarcoma virus
(ASV), and we offer our current view of the mechanisms involved.

The Basic Methodology

The experiments described here utilized either Peking duck embryo
fibroblasts or quail tumor cells (QT-6 cells, derived from a methyl-
cholanthrene-induced fibrosarcoma in Japanese quail) as permissive
hosts for infection by high-titer stocks of B77 strain of ASV; these
stocks also contain high titers of transformation-defection deletion
mutants of ASV. Because the amount of viral DNA synthesized in infected
cells is small (generally from 1-20 copies per cell), analyses require
the high sensitivity provided by annealing of labeled virus-specific
reagents to unlabeled viral DNA present in cell extracts; labeled viral
RNA [(+)strand] is employed to test for (-) strand DNA, and labeled
DNA complementary to the viral genome [cDNA, (-)strand], synthesized
in vitro from a template of viral RNA, is used to test for (+) strand
DNA. Hybridization is usually assessed by resistance of the labeled
reagents to digestion by single-strand specific nucleases.

Cytoplasmic Synthesis of Viral DNA

During the first 5-8 h after infection of avian cells by ASV, viral
DNA is synthesized exclusively in the cytoplasm, as demonstrated by
analysis of fractionated cells or by infection of enucleated cells
(Varmus et al., 1974). Studies with conditional mutants of ASV have
documented that viral DNA synthesis is catalyzed by the RNA-directed
DNA polymerase associated with virus particles (Verma et al., 1976).
The principal form of viral DNA in the cytoplasm is a linear duplex
with a (-) strand the length of a subunit of the viral genome (3×10^6
daltons) and segmented (+) strands of $0.5 - 1.0 \times 10^6$ daltons; shorter
duplexes are also observed and may be either precursors to the full-
length forms or abortive products. Viral DNA can be purified from the
cytoplasm of infected cells by labeling both strands with 5-bromode-
oxyuridine (BUdR) and selecting DNA that bands appropriately in cesium
chloride density gradients and in rate zonal gradients of sucrose.

When observed directly in the electron microscope or by electrophoresis in agarose gels with molecular hybridization, these molecules appear to be linear duplexes of about $5 - 6 \times 10^6$ daltons; mapping of sites for cleavage by restriction endonucleases confirms that only linear duplexes, not open or closed circular duplexes, are recovered from the cytoplasm (Shank et al., manuscript in preparation).

Form I DNA in the Nucleus

After synthesis in the cytoplasm, viral DNA appears in the nucleus, where a portion (up to 50%) is converted to covalently closed circular (form I) duplexes (Guntaka et al., 1975; 1976). The claim that the cytoplasmic DNA is a precursor to form I DNA is supported by experiments in which the cytoplasmic linear forms were labeled with BUdR and then converted into nuclear form I DNA during a "chase" with thymidine (Shank and Varmus, submitted for publ., 1977). The mechanism of conversion from linear to circular forms is unknown, as discussed below.

A variety of physicochemical procedures have been employed to purify form I DNA from the nuclei of infected QT-6 cells (Guntaka et al., 1976). When the size of form I DNA was analyzed by electron microscopy or by electrophoresis in agarose gels, three size classes of circular viral DNA were observed: DNA of ca. 6.6×10^6 daltons, presumed to be synthesized from a subunit of the ASV genome; DNA of ca. 5.6×10^6 daltons, presumed to be synthesized from a subunit of the genome of transformation-defective deletion mutants known to be present in the infecting stocks; and DNA of ca. $2 - 4 \times 10^6$ daltons, presumed to represent highly defective molecules generated by errors during DNA synthesis. As expected, these small circles are not infectious under conditions that allow successful transfection by the two larger species. The genetic composition of the small circles, however, is not known, and it is not known whether they can integrate into the host cell genome.

Steps in Synthesis of Viral DNA (cf. Figure 1)

Based upon these observations and upon studies of the behavior of viral DNA polymerase in vitro in this laboratory and others, it is possible to propose five sequential steps in the synthesis of viral DNA: (1) *initiation of (-) strand synthesis,* using the tRNA[trp], which is bound to the RNA genome 101 bases from the 5' end, as primer (Taylor, 1977) and viral genome as template; (2) a *"transcriptional leap,"* which allows transcription to proceed beyond the 5' end, at which it is presumably stopped, to the 3' end of the same or a different subunit; (3) *initiation of (+) strand synthesis,* possibly using remnants of viral RNA partially digested by RNase H as primer and (-) strand DNA as template; (4) *completion of the DNA strands,* presumably by continuous synthesis of the (-) strand and multiple initiations of the segmented (+) strands; and (5) *circularization* of the linear duplex *and ligation* of all ends to form a covalently closed circle of duplex DNA (form I). All of these steps, save the circularization and ligation, are likely to occur in the cytoplasm of infected cells and may require no other enzymes than the viral DNA polymerase with its associated RNase H activity.

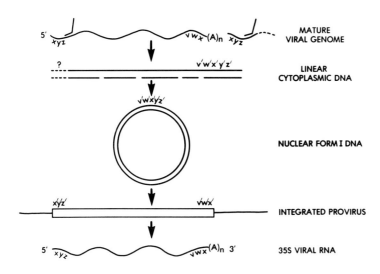

Fig. 1. Provisional model for the synthesis, integration, and expression of virus-specific DNA. For discussion, see text

Recent studies have confirmed some aspects of this scheme and have also created new difficulties to be resolved:

1. Mapping of linear DNA from infected cells with restriction endonucleases and analysis with hybridization reagents specific for small regions of the genome thus far support the notion that synthesis begins, in vivo as well as in vitro, near the 5' end of the genome, presumably primed by tRNA^{trp} (unpublished results of the authors).

2. DNA sequencing methods have been applied to transcripts from the 3' and 5' termini of the viral genome, revealing a natural repeat of 16 - 21 bases (termed "x" in Fig. 1) at the ends (Haseltine et al., 1977; Schwartz et al., 1977; Shine et al., 1977). This suggests a mechanism, as yet unproved, by which the "transcriptional leap" can be effected: after synthesis of the DNA complement of the 5' terminus, the RNA in this hybrid is either displaced or degraded, permitting the DNA to base-pair with the terminal redundancy on the 3' side and thereby prime further synthesis. One obvious limitation of this maneuver is the loss of the redundancy, which must be recovered at some later step to permit the synthesis of competent viral RNA by transcription of proviral DNA. Since "x" would appear to be the site on the viral genome at which "proper" integration occurs (see Fig. 1), a staggered endonucleolytic cleavage at this site during integration could restore the redundancy. However, other solutions - including tandem integration or integration at an "x" sequence present in the normal cellular DNA - have not been excluded.

3. Although one relatively homogeneous class of "plus" strands has been identified in DNA synthesized in vivo (a piece of ca. 300 nucleotides containing linked sequences from both 3' and 5' termini; unpublished data of authors), most of the pieces are extremely heterogeneous in size. There is currently no direct information about their primers, and it is not known whether they initiate and terminate at specific sites.

4. The rate of elongation of viral DNA appears to be extremely slow;
in ASV-infected cells, 2 to 3 h are required to complete "minus" strands
of 10,000 bases, suggesting a polymerization rate of 0.5 to 1 nucleotide
per second (unpublished data of authors). The explanation for this
slow rate is unknown, but in cultured cells rendered stationary by
serum deprivation, the rate is at least ten-fold lower (Varmus et al.,
1977). The concentration of deoxynucleotide substrates or the avail-
ability of nucleic acid binding proteins could be influencing the rates
of synthesis.

5. Circularization could occur by homologous recombination between
sequences redundant in linear DNA (e.g. "xyz" in Fig. 1) or by synthesis
of complementary single-stranded tails ("sticky ends"), which would
have to involve the binding site for tRNA^{trp}. However, it is not yet
possible to choose between these alternatives; it is not even known
whether the 5' terminal sequence is present twice in the linear molecule (as denoted by the question mark). In either case, however,
there is no orthodox fashion by which the mechanism of circularization
could reconstitute the natural repetition presumably lost during the
"transcriptional leap."

Integration of Viral DNA

Circumstantial evidence (Guntaka et al., 1975) intuition, and the
analogy with other animal and bacterial viruses suggest that the
form I viral DNA is the form that integrates into the host chromosome.
Covalent integration of up to several copies of viral DNA has been
documented with several techniques (14, 15). However, the number and
nature of sites in the host genome at which integration may occur are
unknown, as is the mechanism of integration.

The process of integration of RNA tumor virus genomes into the chromosomes of permissive cells is likely to require a higher degree of
specificity than appears to exist for the integration of SV40 DNA
into the DNA of nonpermissive cells (Botchan et al., 1976; Ketner and
Kelly, 1976). To obtain a functional provirus from which mature viral
genome and viral mRNAs can be transcribed, the viral DNA must be properly oriented, and the terminal redundancy ("x") must be restored
(Fig. 1). In addition, there may be a requirement for integration
"downstream" from an active cellular promoter for RNA polymerase;
otherwise the high rate of synthesis of viral RNA observed in these
cells would be dependent upon a viral promoter that must itself be
copied into RNA. In the case of the mouse mammary tumor virus, which
is under transcriptional control by glucocorticoids even after infecting heterologous cells (Ringold et al., 1977; Varmus et al., 1978),
there may be a special requirement for integration near a steroid-
responsive site in host DNA. Thus, we would predict that at least one
provirus in each virus-producing cell will be integrated by recombination between a preferred site in the cell DNA and a unique site in
viral DNA. Mapping of integrated DNA with restriction endonucleases
(Botchan et al., 1976) should soon permit a test of this hypothesis.

Unintegrated DNA in Chronically Infected Cells

Generally, we find 2-8 copies of viral DNA integrated into the genome
of permissive (avian) cells (Varmus et al., 1975), but not all of the

viral DNA synthesized early after infection is integrated (Varmus et
al., 1976). The failure to integrate could be due to limitation of
integration sites or to some defect in the synthesized DNA. In addition
to these early species of unintegrated DNA, however, we have observed
recently synthesized, unintegrated viral DNA in *chronically* infected
cells. In ASV-infected duck embryo fibroblasts, viral DNA is synthesized
in the cytoplasm several weeks after the initial infection (Varmus and
Shank, 1976). The structure of the DNA [linear duplexes of subunit-
length (-) strand and segmented (+) strands] and the mode of incorporation of density label into the DNA suggest that it is made by viral
reverse transcriptase, perhaps in nascent virus particles in the
cytoplasm of these virus-producing cells. In ASV-infected QT-6 cells,
on the other hand, we find form I viral DNA persisting in the nucleus
in an unintegrated state up to 600 h after infection (Guntaka et al.,
1976).

Although the forms and location of free viral DNA may differ, it is
likely that all cells chronically infected with RNA tumor viruses
may contain unintegrated as well as integrated DNA. For example, we
recently observed unintegrated viral DNA in rat hepatoma cells and
mink lung cells chronically infected with murine mammary tumor virus
(Ringold et al., 1977); unpublished data of C. Cohen and authors).
There is, however, as yet no evidence that unintegrated DNA serves as
a template for synthesis of viral RNA or that it serves any other
function in the virus life cycle.

Acknowledgements. The work was supported by grants from the American
Cancer Society and the National Cancer Institute. H.E.V. is a recipient
of a Career Development Award from the USPHS.

References
===

Botchan, M., Topp, W., Sambrook, J.: Cell 9, 269 (1976)
Guntaka, R.V., Mahy, B., Bishop, J.M., Varmus, H.E.: Nature (London) 253, 507-511
    (1975)
Guntaka, R.V., Richards, O.C., Shank, P.R., Kung, H.J., Davidson, N., Fritsch, E.,
    Bishop, J.M., Varmus, H.E.: J. Mol. Biol. 106, 337 (1976)
Haseltine, W.A., Maxam, A.M., Gibbert, W.: Proc. Natl. Acad. Sci. USA 74, 989 (1977)
Ketner, G., Kelly, T.: Proc. Natl. Acad. Sci. USA 73, 1102 (1976)
Ringold, G.R., Cardiff, R.D., Varmus, H.E., Yamamoto, K.R.: Cell 10, 11 (1977)
Ringold, G.R., Yamamoto, K.R., Shank, P.R., Varmus, H.E.: Cell 10, 19 (1977)
Schwartz, D.E., Zamecnik, P.C., Weith, H.L.: Proc. Natl. Acad. Sci. USA 74, 994
    (1977)
Shine, J., Czernilofsky, A.P., Friedrich, R., Bishop, J.M., Goodman, H.M.: Proc.
    Natl. Acad. Sci. USA 74, 1473 (1977)
Taylor, J.M.: Biochim. Biophys. Acta 473, 57 (1977)
Varmus, H.E., Guntaka, R.V., Deng, C.T., Bishop, J.M.: Cold Spring Harbor Symp.
    Quant. Biol 39, 987 (1975)
Varmus, H.E., Guntaka, R.V., Fan, W., Heasley, S., Bishop, J.M.: Proc. Natl. Acad.
    Sci. USA 71, 3874 (1974)
Varmus, H.E., Heasley, S., Linn, J., Wheeler, K.: J. Virol. 18, 574 (1976)
Varmus, H.E., Padgett, T., Heasley, S., Simon, G., Bishop, J.M.: Cell, 11, 307
    (1977)

Varmus, H.E., Ringold, G.R., Yamamoto, K.R.: In: Glucocorticoid Hormone Action.
    Baxter, J., Rousseau, G. (eds.). Berlin-Heidelberg-New York: Springer Verlag,
    1978, in press
Varmus, H.E., Shank, P.R.: J. Virol. 18, 567 (1976)
Varmus, H.E., Vogt, P.K., Bishop, J.M.: Proc. Natl. Acad. Sci. USA 70, 3067 (1973)
Verma, I.M., Varmus, H.E., Hunter, H.E.: Virology 75, 16 (1976)
Weinberg, R.A.: Biochim. Biophys. Acta 473, 39 (1977)

RNA-dependent DNA Polymerase in a "Virus-Free" System

G. Bauer and P. H. Hofschneider

RNA-dependent DNA polymerase (reverse transcriptase) is known to be a regular constituent of RNA tumor viruses and of relatives of this viral group. The role of the enzyme during replication of these viruses has been clarified (1; see also H. Varmus, this Volume). The viral reverse transcriptase synthesizes a double-stranded DNA copy of the viral RNA genome. This DNA (the "provirus") is subsequently integrated into the host genome and transcribed like a cellular gene.

In 1971 Temin postulated in his "protovirus theory" that RNA-dependent DNA synthesis might not be a unique event occurring during the replication of RNA tumor viruses, but might be of general biologic importance (2). According to his speculations, physiologic reverse transcription of cellular messenger RNA and subsequent recombination and integration events might provide a mechanism for amplification, translocation, and mutation of genetic information.

Until now, however, no enzyme with the ability to transcribe heteropolymeric RNA in vitro has been isolated from a "virus-free" system, i.e., a system not replicating exogenous virus, not expressing endogenous viruses, and being not transformed. This is a report on the isolation and characterization of an ubiquitously appearing RNA-dependent DNA polymerase from embryonated, virus-free chicken eggs. The enzyme is not the gene product of the polymerase gene of known endogenous or exogenous avian RNA tumor viruses: Its expression is independent of the genetic system of the endogenous avian leukosis virus and seems to be regulated by its own genetic system.

The enzyme is found enclosed in sedimentable structures, surrounded by a lipoprotein membrane. These structures have been preliminarily termed "particles." They do not exhibit any of the biologic properties typical of RNA tumor viruses, such as infectivity, helper activity, or interference. Their protein composition is different from that of any of the known avian RNA tumor viruses. The sedimentation constant of the particles is about 400 to 500 S, compared to a value of 680 S for RNA tumor viruses. In the electron microscope the particles appear as a heterogenous population with a size distribution between 20 and 200 nm. The purified particles do not exhibit endogenous DNA synthesis. It is not clear, therefore, whether they carry any nucleic acid. In any case the particles are different from the known avian RNA tumor viruses and seem to be vesicles, released from cells. Release of non-viral particles from cells has also been reported by other groups (3, 4); however, the particles did not carry reverse transcriptase. Using affinity chromatography, an RNA-dependent DNA polymerase has been isolated from purified particles. The enzyme reaction fulfills all the necessary enzymologic criteria to allow quantitative analysis. The enzyme does not work with template or primer alone. The enzyme utilizes homopolymeric template-primer complexes such as poly (C)(dG)$_{12}$, poly (dC) (dG)$_{12}$, poly (A)(dT)$_{12}$ and is able to transcribe the heteropolymeric part of globin mRNA. Analysis of the globin mRNA-directed reaction product, using single-strand specific nuclease S$_1$ and Cs$_2$SO$_4$ density gradient analysis, has shown that the DNA synthesized by the particle enzyme is a specific copy of the template. All these reaction

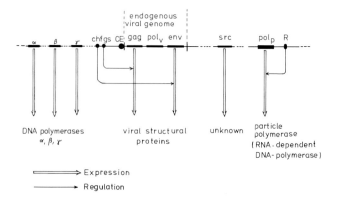

Fig. 1. Polymerase genes and endogenous viral genes in the DNA of normal chicken cells. The DNA contains the genes for DNA polymerases α, β, γ (5, 6); functional gene products can be isolated from the cells. DNA polymerases α and β are classical DNA-dependent DNA polymerases, whereas polymerase γ is characterized by its ability to synthesize poly (dT) dependent on a poly (rA) template. However, the enzyme can not use other homopolymeric or heteropolymeric RNA templates and therefore can be easily distinguished from "true" RNA-dependent DNA polymerase, which can transcribe heteropolymeric RNA (6).

The part concerning the endogenous viral genome is a summary of a lot of relevant literature (for review see 10, 11). It has been shown that the chicken genome contains the genome of the avian leukosis virus. The complete expression of this genetic information, leading to the formation of replicative virus, is prevented by a control mechanism, which has been suggested to be a cis-acting control element (12). This control can be overcome by application of chemicals or radiation (13). Some lines of chickens spontaneously express the complete sequence. The result is the appearance of a replicative, nonpathogenic virus, which is called RAV - O (14). Whereas the endogenous ALV information is located on macrochromosomes, microchromosomes contain a sequence closely related to the transforming gene of avian sarcoma viruses (15-17). This gene is transcribed in normal cells; however, the gene product is not known.

In normal chicken cells the genes *gag* and *env* may or may not be expressed. The expression is regulated by regulatory genes called *gs* and *chf*, which have the alleic forms gs^- or gs^+ and chf^- or chf^+. The alleles are inherited in a Mendelian way, whereas the structural genes *gag* (codes for the precursor of the group specific antigens) and *env* (codes for envelope glycoprotein) are present in the same form in all chicken cells. The gene for the viral RNA-dependent DNA polymerase (pol_v) is present in the genome of the chicken cell. The presence of the gene is demonstrated by the fact that RAV - O, which is induced from cells, contains functional RNA-dependent DNA polymerase. [The polymerase of the endogenous virus is closely related to or identical with the polymerase of exogenous viruses of the ALV/ASV group (9; own unpublished experiments).]

However, the polymerase gene of the endogenous viral genome is not expressed in normal, non(-)virus-producing cells, because the product of this gene could not be demonstrated in normal cells (18), and because mutant exogenous viruses with a defect in their gene "pol" are not complemented by normal chicken cells (19). From the data summarized here and published elsewhere (20-23) we suggest the presence of a second gene for RNA-dependent DNA polymerase in the genome of normal chicken cells (pol_p). The product of this gene is a functional RNA-dependent DNA polymerase with the ability to transcribe heteropolymeric and homopolymeric RNA into complementary DNA. The enzyme is different from the polymerases of the known exogenous and endogenous RNA tumor viruses of chicken. Therefore the gene pol_p is different from the gene pol_v of the endogenous ALV. The expression of the gene pol_p is independent of the system regulating the complete or partial expression of the endogenous viral genome.

properties show that the enzyme is different from the DNA polymerases
α, β, γ (5-7) and from terminal deoxynucleotidyl transferase and must
be regarded as a true RNA-dependent DNA polymerase.

The next step was to determine whether the particle enzyme is the gene
product of the polymerase gene of any known chicken RNA tumor virus.
Comparison of the particle enzyme to the polymerase of the reticuloendotheliosis virus (REV) group showed that the two enzymes differ
markedly in their sedimentation constants in glycerol gradients and
in their preference for specific divalent cations. Comparison of the
particle enzyme to the polymerase of the avian leukosis virus/avian
sarcoma virus (ALV/ASV) group was of special interest, because the
genetic information of this virus group is contained in the genome of
normal chicken cells. As can be seen from Figure 1, the complete expression of the information as well as the expression of viral structural components is under genetic control. The question was whether the
particle enzyme was coded by the endogenous viral polymerase gene or
by a different gene. Since the polymerase of the ALV/ASV group is
group-specific whithin the group (8, 9), we used the polymerase of
avian mucosa virus (AMV) as a representative polymerase. In the IgG
inhibition test, using IgG directed against AMV polymerase, particle
polymerase is only weakly inactivated. A 40- to 240-fold excess of IgG,
compared to the homologous enzyme, has to be applied to obtain 50%
inhibition. This demonstrates that the particle enzyme is not identical
to the ALV/ASV polymerase. It remains to be established whether the
weak cross-reaction is specific and due to a relationship of both
enzymes. For further characterization, a biochemical comparison of
particle enzyme to AMV enzyme was performed. Both enzymes differ in
their optima for Mg^+ and K^+ and in various characteristics such as
mode of reaction to increasing concentrations of substrate and thermal
stability. Interestingly, the particle enzyme cannot be protected from
heat inactivation by the presence of viral RNA, whereas the viral
enzyme is essentially stabilized by its natural template. Together,
the immunologic and enzymologic data indicate that the particle enzyme
is not the gene product of endogenous viruses of the ALV/ASV group.

The particle polymerase has been found in the eggs of all flocks of
chicken tested so far. Within one flock, each egg tested was positive.
The presence of activity in eggs from chickens with the genotype
gs⁻chf⁻ shows that the expression of the particle polymerase gene is
independent of the regulatory system controlling the partial expression
of the endogenous ALV.

Fig. 1 (continued)
At least at a certain time during embryogenesis the gene pol_p is expressed in the
embryos of all lines of chickens tested. The expression is obviously controlled by
a regulatory system, whose mechanism and significance remain to be clarified.

α, β, γ: genes for DNA polymerases α, β, γ;
chf, gs: regulatory genes, controlling the expression of structural proteins of
endogenous avian leukosis virus
CE: cis-acting control element
gag: gene for viral structural components
pol_v: gene for viral RNA-dependent DNA polymerase
env: gene for envelope proteins
src: DNA that is homologous to the "transforming" gene of avian sarcoma viruses
pol_p: gene for particle polymerase
R: postulated regulatory element controlling the activity of the gene pol_p

Our data (summarized in Table 1) show that in the chicken system, RNA-dependent DNA polymerase is present in a normal, nontransformed system, without replication of exogenous RNA tumor viruses or the partial or complete expression of the endogenous viral genome. Thus, during embryogenesis a genetic system that is different from and independent of the endogenous ALV is active, providing a functional reverse transcriptase (Fig. 1). It cannot be excluded for the moment that it reflects the presence of a still unknown virus system in chickens. However, the regular occurrence of the activity in embryonated eggs suggests that the enzyme might have a physiologic function. Whether the enzyme is involved in information transfer, gene transposition, or amplification in the cell remains to be established. It will also be of interest to determine whether the release of particle-bound enzyme from cells is artificial or functional.

Table 1. RNA-dependent DNA polymerase from uninfected chicken eggs

---

A) Occurrence:

Ubiquitously appearing activity in the allantoic and amniotic fluid of embryonated chicken eggs, including those from parents with the genotype gs⁻ chf⁻.

The enzyme is enclosed in sedimentable structures (particles). Particles are heterogeneous in size, not infectious, do not show helper activity or interference, do not exhibit endogenous DNA synthesis, and differ in their protein composition from known avian RNA tumor viruses.

B) Purification:

Purification of the particles by centrifugation. The enzyme is subsequently purifed from the particles by affinity chromatography.

C) General Properties:

RNA- and DNA-dependent DNA polymerase activity. The enzyme needs template, base-paired primer, and deoxynucleoside triphosphates. The reaction is linear with respect to time, concentration of enzyme, and concentration of template-primer complex. The enzyme can use homopolymeric template-primer complexes such as poly (C) $(dG)_{12}$, poly (dC) $(dG)_{12}$, or poly (A) $(dT)_{12}$. It can transcribe the heteropolymeric part of globin mRNA $[+(dT)_{12}]$ or of 70 S RNA from AMV $[+(dT)_{12}]$.

D) Distinction of the Particle Polymerase from the Polymerases of Known Chicken RNA Tumor Viruses:

Distinction from the polymerase of the reticuloendotheliosis virus group by size and preference for Mg^{2+}.

Distinction from the polymerase of the ALV/ASV group immunologically and enzymologically. Weak inactivation of particle enzyme by IgG directed against the polymerase of AMV. Particle enzyme and enzyme of the ALV/ASV group differ in their thermal stability, mode of response to increasing concentrations of dGTP, optimum curves for Mg^{2+} and K^+. Whereas the viral enzyme can be protected from heat inactivation by viral RNA, this RNA has no stabilizing effect on particle polymerase.

E) Function:

Not yet known. Possibly interesting.

Experiments in progress show that the enzyme concentration goes through a distinct peak during embryogenesis. These results suggest that the enzyme may play a role during embryogenesis.

Acknowledgements. We thank Dr. R.R. Friis (Universität Gießen) for performing biologic control experiments. The excellent technical assistance by Mrs. G. Jilek is appreciated. This work was supported by the Deutsche Forschungsgemeinschaft (SFB 51).

References

1. Verma, I.M.: Biochim. Biophys. Acta 473, 1-38 (1977)
2. Temin, H.M.: J. Natl. Cancer Inst. 46, III-VII (1971)
3. Margalith, M., Gerard, G.F., Green, M.: Biochim. Biophys. Acta 425, 305-315 (1976)
4. Smith, C.C., Maverakis, N.H., Ackermann, W.W.: Proc. Soc. Exp. Biol. Med. 152, 645-650 (1976)
5. Brun, G., Rougeon, F., Lamber, M., Chapeville, F.: Eur. J. Biochem. 41, 241-251 (1974)
6. Bolden, A., Fry, M., Muller, R., Citarella, R., Weissbach, A.: ABB 153, 26-33 (1972)
7. Weissbach, A.: Cell 5, 101-108 (1975)
8. Nowinski, R.C., Watson, K.F., Yaniv, A., Spiegelman, S.: J. Virol. 10, 959-964 (1972)
9. Mizutani, S., Temin, H.M.: J. Virol. 12, 440-448 (1973)
10. Tooze, J. (ed.): The Molecular Biology of Tumour Viruses. New York: Cold Spring Harbor Laboratory 1973
11. Temin, H.M.: Adv. Cancer Res. 19, 47-104 (1974)
12. Cooper, G.M., Temin, H.M.: J. Virol. 17, 422-430 (1976)
13. Weiss, R.A., Friis, R.R., Katz, E., Vogt, P.K.: Virology 46, 920-928 (1971)
14. Vogt, P.K., Friis, R.R.: Virology 43, 223-234 (1971)
15. Stehelin, D., Varmus, H.E., Bishop, J.M., Vogt, P.K.: Nature (London) 260, 170-173 (1976)
16. Varmus, H.E., Stehelin, D., Spector, D., Tal, J., Fujita, D., Padgett, T., Roulland - Dussoix, D., Kung, H.-J., Bishop, J.M.: In: Animal Virology. Baltimore, D. et al. (eds.). 1976, pp. 339-358
17. Padgett, T.G., Stubblefield, E., Varmus, H.E.: Cell 10, 649-657 (1977)
18. Weissbach, A., Bolden, A., Muller, R., Hanafusa, H., Hanafusa, T.: J. Virol. 10, 321-327 (1972)
19. Mason, W.S., Friis, R.R., Lineal, M., Vogt, P.K.: Virology 61, 559-574 (1974)
20. Bauer, G., Hofschneider, P.H.: Proc. Natl. Acad. Sci. USA 73, 3025-3029 (1976)
21. Bauer, G., Jilek, G., Hofschneider, P.H.: In: Modern Trends in Human Leukemia II. Neth, R., et al. (eds.). München: J.F. Lehmanns Verlag, 1976, pp. 515-530
22. Bauer, G.: Thesis. Universität München 1976
23. Bauer, G., Jilek, G., Hofschneider, P.H.: J. Virol. in press (1977)

Studies on the Cellular Mechanisms of Chemical Oncogenesis

C. Heidelberger

It is now generally recognized and widely accepted that environmental
chemicals are a major cause of human cancer. This conclusion has been
reached on the basis of extensive epidemiologic data and tragic in-
dustrial exposures (Higginson and Muir, 1973). Thus, it is urgently
necessary to increase our understanding of the fundamental cellular
and molecular mechanisms of chemical carcinogenesis, as well as to
develop valid in vitro screening tests for the surveillance of the
environment. I will address myself to both of these needs.

One of the most important principles that has been learned is that most
chemical carcinogens are not by themselves the true carcinogen, but
require metabolic activation to proximal or ultimately carcinogenic
forms. Despite the bewildering variety of the structures of chemical
carcinogens, it has now been shown that without exception their activated forms are electrophilic in nature and react with cellular macromolecular nucleophiles to form covalent bonds (Miller, 1970). The
enzyme systems that carry out this metabolic activation of almost all
carcinogens are the cytochrome P450 mixed-function oxidases, which
are inducible and are ordinarily considered to be detoxication enzymes. It is, therefore, an irony of nature that these same enzymes
activate carcinogens to their noxious form (Heidelberger, 1975). It
is now generally accepted that the activated form of carcinogens ini-
tiates the process of carcinogenesis as a result of the covalent binding
to a cellular macromolecular target. There is increasing evidence that
this target is DNA, but as yet there is no proof that this is so.
Polycyclic aromatic hydrocarbons (PAH), such as benzo[a]pyrene (BP),
are ubiquitous air pollutants, and recent work has demonstrated that
BP is activated to a diol epoxide, which is formed stereospecifically,
and which is bound to DNA and RNA through the 2-amino group of guanine
(Sims et al., 1974; Weinstein et al., 1976; Yang et al., 1976).

Let me now formulate three fundamental questions on the cellular mechanisms of chemical oncogenesis: 1) Does the chemical transform normal
cells to cancer cells, or does it select for preexisting cancer cells?;
2) Does the chemical transform cells by itself, or does it act through
the "switch-on" of an oncogenic virus?; and 3) If the chemical trans-
forms cells by itself, is the mechanism mutational or nonmutational;
the latter would be some perpetuated derepression of genetic information.

To answer these questions, the second one being most pertinent to this
Symposium, it was necessary to develop cell culture systems in which
chemical carcinogens could produce oncogenic transformation. Two types
of systems have been developed. The first is a cloning assay using
primary or secondary cultures of hamster embryo cells, which was pio-
neered by Berwald and Sachs (1965) and extended by DiPaolo et al. (cf.
1969). These cells are diploid, clone with a low efficiency, and can
be scored for transformation within 8-10 days. The second type of
system involves the use of permanent lines of mouse or other rodent
fibroblasts, one of which from mouse prostate and one from mouse
embryo fibroblasts were developed in our laboratory (Chen and Heidelberger, 1969; Reznikoff et al., 1973a, b). I have discussed in detail
the relative advantages and disadvantages of each of these systems

(Heidelberger, 1973a, b; 1975), but will devote the most attention to our own research. With both of these systems the nontransformed cells exhibit density-dependent inhibition of cell division and form flat clones in the dishes, whereas the transformed cells lose that inhibition, and pile up to form thick criss-cross patterned clones, which in almost all cases give rise to fibrosarcomas on inoculation into syngeneic hosts. Thus, morphologic and oncogenic transformation are closely related. In all these systems, PAH produce dose-dependent transformation with a frequency that is proportional to the carcinogenic activities of the compounds in vivo (Berwald and Sachs, 1965; DiPaolo et al., 1969; Chen and Heidelberger, 1969; Reznikoff et al., 1973b).

Using the lines of mouse fibroblasts, we showed early in the game that epoxides of PAH were much more active than the parent hydrocarbons or their other metabolites at producing oncogenic transformation (Grover et al., 1971), covalent binding to DNA, RNA, and proteins (Kuroki et al., 1971/72), and induction of 8-azaguanine-resistant mutants in Chinese hamster V79 cells (Huberman et al., 1971). Moreover, we showed that induction of the mixed-function oxidases increased, and inhibition thereof decreased the transformation produced by 3-methylcholanthrene (MCA) (Marquardt and Heidelberger, 1972; Nesnow and Heidelberger, 1976). This proves that mixed-function oxidase-mediated activation of PAH is required for oncogenic transformation and chemical mutagenesis.

Soon after the development of a quantitative system, we were able to transform single cells in individual dishes with a very high efficiency, which ruled out the possibility that the chemicals were selecting for preexisting transformed cells (Mondal and Heidelberger, 1970). This conclusion has been reached in several other laboratories; hence, the first question has been answered by the use of cell cultures, as it could not have been in vivo.

I shall return to the second question below, and will address the third one: If there are situations in which a chemical can transform cells without the intervention of an oncogenic virus, is the mechanism mutational or nonmutational? Formerly, it was thought that there was an inverse relationship between mutagenic and carcinogenic chemicals. However, that was before the generality of metabolic activation had been established, and the microorganisms that were used for mutagenesis testing lacked the mixed-function oxidases required to activate the precarcinogen. Subsequently, when activated forms of carcinogens have been tested, they have almost invariably been found to be mutagenic, as we have shown in bacteriophage T4 (Corbett et al., 1970) and mammalian cells (Huberman et al., 1971). Moreover, in the well-known test developed by Ames, compounds are incubated with a microsomal preparation from rat liver and produce reversion mutations in suitable tester strains of *Salmonella typhimurium*. A startlingly good correlation was obtained between carcinogenic activities in vivo and the mutations produced in this bacterial system with a series of 300 compounds (McCann et al., 1975). This system is being universally adopted as a prescreen for environmental mutagens and by implication, carcinogens. With the simple-minded belief that *Salmonella* do not get cancer, we considered it worthwhile to develop an analog of the Ames system in which mammalian cells that lack mixed-function oxidase activity were incubated with rat liver microsomal preparations; mutagenicity at the HGPRT locus was obtained with PAH and with aflatoxins (Krahn and Heidelberger, 1977). We are trying to extend the work to other classes of chemical carcinogens, in the belief that such a system of mammalian cell mutagenesis, although more expensive and time-consuming than the Ames test, might find a place in a tier system for environmental prescreening.

With a model system such as our mouse fibroblast C3H/10T1/2 cells, it is continually necessary to validate it by comparison of various properties with the in vivo carcinogenesis situation. One of the most sophisticated and least understood properties of chemically induced tumors is that they usually have individual and not cross-reacting cell-surface tumor-specific transplantation antigens (TSTA). We showed in our cell cultures that multiple chemically transformed clones derived from a highly cloned nontransformed line, had individual and not cross-reacting TSTAs (Mondal et al., 1970; Embleton and Heidelberger, 1972). This shows that the TSTA is produced during the process of transformation and is not selected for out of the parental population. Moreover, we found that on the surface of the same chemically transformed clones there were individual TSTAs and common embryonic antigens (Embleton and Heidelberger, 1975). Thus, the model system behaves properly with respect to surface antigens. We have also demonstrated in these cells that transformation by the short-acting alkylating agent, MNNG, is cell cycle phase-specific, the sensitive period being about 4 h prior to the onset of DNA synthesis (Bertram and Heidelberger, 1974).

The classical research of Berenblum and his colleagues showed that carcinogenesis in mouse skin can be divided into two phases: an irreversible process known as initiation, and a reversible process termed promotion. PAH are very effective initiators, and croton oil and its active constituent phorbol esters are promoters (Berenblum, 1954). We have now demonstrated that two-stage transformation occurs in our C3H/10T1/2 mouse embryo fibroblasts, which is in many ways exactly analogous to the situation in mouse skin (Mondal et al., 1976). In this system, ultraviolet light acts as a pure initiator (Mondal and Heidelberger, 1976). The tools are now available to investigate separately the cellular and molecular mechanisms of initiation and promotion, as well as to prescreen for environmental promoters, which may be just as important in the induction of human cancer as are complete carcinogens. Such studies are underway in our laboratory.

Before returning specifically to question 2, I will review briefly some of the interactions between chemical carcinogens and oncogenic viruses. A considerable literature of work prior to 1960 in which pox virus-treated chickens and influenza virus-treated mice were more susceptible to chemical carcinogenesis than uninfected animals was reviewed by Duran-Reynals (1963) and Martin (1964). Since oncogenic viruses were not involved in those studies it is difficult to draw any firm conclusions.

In 1963, Irino et al. (1963) discovered that application of MCA to the skin of RF mice induced leukemias, and a cell-free filtrate obtained from several of the leukemic mice gave rise to a few leukemias following inoculation into recipient mice; viral particles were seen in the lymph nodes of such recipients. More recently Basombrio (1973) produced leukemias in BALB/c mice by cell-free extracts of MCA-induced sarcomas. This indicates that a mouse leukemia virus (MuLV) was activated in the process of carcinogenesis with these chemicals, a situation rather analogous to the radiation-induced leukemia virus (DeCleve et al., 1975). Ball and McCarter (1971) demonstrated that a single neonatal injection of 7,12-dimethylbenz[a]anthracene (DMBA) into CFW/D mice increased the level of MuLV in the tissues of these mice, and they obtained lymphomas from the cell-free passage of DMBA-induced lymphomas. At the same time that those studies demonstrating cell-free induction of leukemias were being done, considerable research was carried out in which, by complement-fixation techniques, the group-specific (gs)antigens of MuLV were detected in PAH-induced sarcomas in several strains of mice (Igel et al., 1969; Whitmire et al., 1971; Whitmire and Salerno,

1972). GS antigens and type C particles were also found in MCA-induced hamster tumors (Freeman et al., 1974). These results were interpreted as supporting the "oncogene" hypothesis (Huebner and Todaro, 1969; Todaro and Huebner, 1972). One of the tenets of that hypothesis is that in most situations chemical carcinogens induce tumors by "switching on" oncogenic viruses or their "oncogenes," which actually are responsible for the oncogenic transformation.

An attempt to find a correlation between the genetic susceptibility of mice to sarcoma induction by MCA and the expression of endogenous MuLV by Kouri et al. (1973) was unsuccessful, and an inverse correlation was observed between endogenous virus expression in various strains of mice and the inducibility of aryl hydrocarbon hydroxylase. In a study of the expression of endogenous MuLV in sarcomas induced by MCA in carefully chosen backcross mice, Nowinski and Miller (1976) concluded that there was no detectable influence of MuLV production on sensitivity to chemical carcinogenesis, in contrast to the effects of MuLV on transformation in cell cultures (see below). Nevertheless, immunization of mice with MuLV decreased the incidence of chemically induced tumors (Whitmire and Huebner, 1972; Whitmire, 1973). It seems justified at present to conclude that there is no strong evidence that endogenous MuLV is causally involved in chemical carcinogenesis and that it is more probably present as a "passenger."

There are some situations in which interactions of DNA viruses and chemical carcinogens have been demonstrated. Casto et al. (1974) showed that chemical carcinogens sharply enhance the transformation of hamster embryo cells induced by adenovirus SA7. They have found some evidence to support the hypothesis that the chemical carcinogens induced damage in the cellular DNA that facilitated integration of the viral genome. Diamond et al. (1974), found that a chemical carcinogen, 4NQO, enhanced SV 40-induced transformation of Chinese hamster embryo cells. To confuse the situation even more, Hatch et al. (1975) found hamster leukemia C-type RNA virus in cells transformed by adenovirus SA7 and chemicals.

There are many illustrations that leukemia viruses affect the transformation of rodent cells by chemical carcinogens. It was found that cultures of rat embryo cells were not morphologically transformed by chemical carcinogens or by the Rauscher leukemia virus. However, productively infected cells were transformed by various chemical carcinogens (Freeman et al., 1970; Price et al., 1972). High passage rat embryo cells that spontaneously express oncornaviruses have been used extensively to study transformation by various chemical carcinogens, and a good correlation has been found with in vivo carcinogenic activity (Freeman et al., 1973; 1975). A similarly good correlation was observed in mouse embryo cells infected with a nontransforming AKR MuLV (Rhim et al., 1971; 1974). Rasheed et al. (1976a, b) found that rat embryo cells become spontaneously transformed many passages after they started to release an endogenous ecotropic virus (RaLV), which is unique in that it propagates in its own cell of origin, and also renders the cells more sensitive to transformation by chemical carcinogens. Finally, it has been found (Price et al., 1977) in rat embryo cells infected by a xenotropic MuLV, that a specific antibody to that virus inhibited oncogenic transformation by MCA. The interpretation given was that the virus produces specific intracellular conditions required for the chemical transformation of these cells.

I will now return to question 2 and how our own research pertains to it. We asked whether there was any "switch on" of MuLV when the C3H/10T1/2 cells underwent oncogenic transformation. Six nontransformed clones, 16 independently chemically transformed clones, and 3 spon-

taneously transformed clones were examined for infectious ecotropic
viruses by the XC test, for the gsl (P30) antigen by a monovalent
antiserum, and for the presence in the medium of RNA-directed DNA
polymerase; all these tests were negative (Rapp et al., 1975). On the
other hand, nontransformed and transformed clones of AKR mouse embryo
fibroblasts were positive in all these tests. These experiments led
to the conclusion that the genotype of the cell determines the expression of endogenous MuLV and its cores (P30 and "reverse transcriptase" are in the viral core) and that the transformed phenotype does
not. Treatment of the C3H cells with 5-iodo-2'-deoxyuridine (IUDR) did
not transform the cells but did induce an endogenous virus, which
further shows that such induction is irrelevant to the transformation
process. Interestingly, the kinetics of induction were different in
the nontransformed and transformed clones. The reason for this is not
yet understood. The question remains to be answered as to whether there
is a "switch on" of viral RNA during the process of transformation.
Information on the expression of the src gene would be particularly
relevant. Such experiments, using appropriate molecular hybridization
probes, are currently in progress. When these results are in, we
should have definitive information on whether oncogenic viruses or
their genes are involved in the transformation produced by chemical
carcinogens in this model system.

In conclusion, considerable progress is being made in elucidating the
cellular mechanisms of chemical carcinogenesis, an understanding of
which must precede a solution of the molecular mechanisms. In the
course of this work, prescreens for environmental carcinogens are
being developed.

References

Ball, J.K., McCarter, J.A.: J. Natl. Cancer Inst. 46, 751-762 (1971)
Basombrio, M.A.: J. Natl. Cancer Inst. 51, 1157-1162 (1973)
Berenblum, I.: Cancer Res. 14, 471-477 (1954)
Bertram, J.S., Heidelberger, C.: Cancer Res. 34, 526-537 (1974)
Berwald, Y., Sachs, L.: J. Natl. Cancer Inst. 35, 641-661 (1965)
Casto, B.C., Pieczynski, W.J., DiPaolo, J.A.: Cancer Res. 34, 72-78 (1974)
Chen, T.T., Heidelberger, C.: Int. J. Cancer 4, 166-178 (1969)
Corbett, T.H., Heidelberger, C., Dove, W.F.: Mol. Pharmacol. 6, 667-679 (1970)
DeCleve, A., Travis, M., Weissman, I.L., Lieberman, M., Kaplan, H.S.: Cancer Res.
    35, 3585-3595 (1975)
Diamond, L., Knorr, R., Shimizu, Y.: Cancer Res. 34, 2599-2604 (1974)
DiPaolo, J.A., Donovan, P.J., Nelson, R.L.: J. Natl. Cancer Inst. 42, 867-874 (1969)
Duran-Reynals, M.L.: Prog. Exp. Tumor Res. 3, 148-185 (1963)
Embleton, M.J., Heidelberger, C.: Int. J. Cancer 9, 8-18 (1972)
Embleton, M.J., Heidelberger, C.: Cancer Res. 35, 2049-2055 (1975)
Freeman, A.E., Igel, H.J., Price, P.J.: In Vitro 2, 107-116 (1975)
Freeman, A.E., Kelloff, G.J., Vernon, M.L., Lane, W.T., Capps, W.I., Baumgarner,
    S.D., Turner, H.C., Huebner, R.J.: J. Natl. Cancer Inst. 52, 1469-1476 (1974)
Freeman, A.E., Price, P.J., Igel, H.J., Young, J.C., Maryak, J.M., Huebner, R.J.:
    J. Natl. Cancer Inst. 44, 65-78 (1970)
Freeman, A.E., Weisburger, E.K., Weisburger, J.H., Wolford, R.G., Maryak, J.M.,
    Huebner, R.J.: J. Natl. Cancer Inst. 51, 799-808 (1973)
Grover, P.L., Sims, P., Huberman, E., Marquardt, H., Kuroki, T., Heidelberger, C.:
    Proc. Natl. Acad. Sci. USA 68, 1098-1101 (1971)
Hatch, G.G., Casto. B.C., McCormick, K.J., and Trentin, J.J.: Cancer Res. 35,
    3792-3797 (1975)
Heidelberger, C.: Fed. Proc. 32, 2154-2161 (1973a)
Heidelberger, C.: Adv. Cancer Res. 18, 317-366 (1973b)

Heidelberger, C.: Annu. Rev. Biochem. 44, 79-121 (1975)
Higginson, J., Muir, C.S.: In: Cancer Medicine. Holland, J.F., Frei, E. (eds.)
    Philadelphia: Lea and Febiger, 1975, pp. 241-306
Huberman, E., Aspiras, L., Heidelberger, C., Grover, P.L., Sims, P.: Proc. Natl.
    Acad. Sci. USA 68, 3195-3199 (1971)
Huebner, R.J., Todaro, G.J.: Proc. Natl. Acad. Sci. USA 64, 1087-1094 (1969)
Igel, H.J., Huebner, R.J., Turner, H.C., Kotin, P., Falk, H.L.: Science 166,
    1636-1639 (1969)
Irino, S., Ota, Z., Sezaki, T., Suzaki, K.: Gann 54, 225-237 (1963)
Kouri, R.E., Ratrie, H., Whitmire, C.E.: J. Natl. Cancer Inst. 51, 197-200 (1973)
Krahn, D.F., Heidelberger, C.: Mutat. Res. 46, 27-44 (1977)
Kuroki, T., Huberman, E., Marquardt, H., Selkirk, J.K., Heidelberger, C., Grover,
    P.L., Sims, P.: Chem.-Biol. Interactions 4, 389-397 (1971/72)
Marquardt, H., Heidelberger, C.: Cancer Res. 32, 721-725 (1972)
Martin, C.M.: Prog. Exp. Tumor Res. 5, 134-156 (1964)
McCann, J., Choi, E., Yamasaki, E., Ames, B.N.: Proc. Natl. Acad. Sci. USA 72,
    5135-5139 (1975)
Miller, J.A.: Cancer Res. 30, 559-576 (1970)
Mondal, S., Brankow, D.W., Heidelberger, C.: Cancer Res. 36, 2254-2260 (1976)
Mondal, S., Heidelberger, C.: Proc. Natl. Acad. Sci. USA 65, 219-225 (1970)
Mondal, S., Heidelberger, C.: Nature (London) 260, 710-711 (1976)
Mondal, S., Iype, P.T., Griesbach, L.M., Heidelberger, C.: Cancer Res. 30 1593-1597
    (1970)
Nesnow, S., Heidelberger, C.: Cancer Res. 36, 1801-1808 (1976)
Nowinski, R.C., Miller, E.C.: J. Natl. Cancer Inst. 57, 1347-1350 (1976)
Price, P.J., Suk, W.A., Freeman, A.E.: Science 177, 1003-1004 (1972)
Price, P.J., Suk, W.A., Peters, R.L., Gilden, R.V., Huebner, R.J.: Proc. Natl. Acad.
    Sci. USA 74, 579-581 (1977)
Rapp, U.R., Nowinski, R.C., Reznikoff, C.A., Heidelberger, C.: Virology 65, 392-409
    (1975)
Rasheed, S., Bruszewski, J., Rongey, R.W., Roy-Burman, P., Charman, H.P., Gardner,
    M.B.: J. Virol. 18, 799-803 (1976a)
Rasheed, S., Freeman, A.E., Gardner, M.B., Huebner, R.J.: J. Virol 18, 776-782
    (1976b)
Reznikoff, C.A., Bertram, J.S., Brankow, D.W., Heidelberger, C.: Cancer Res. 33,
    3239-3249 (1973a)
Reznikoff, C.A., Brankow, D.W., Heidelberger, C.: Cancer Res. 33, 3231-3238 (1973b)
Rhim, J.S., Creasy, B., Huebner, R.J.: Proc. Natl. Acad. Sci. USA 68, 2212-2216
    (1971)
Rhim, J.S., Park, D.K., Weisburger, E.K., Weisburger, J.H.: J. Natl. Cancer Inst.
    52, 1167-1173 (1974)
Sims, P., Grover, P., Swaisland, A., Pal, K., Hewer, A.: Nature (London) 252,326-327
    (1974)
Todaro, G.J., Huebner, R.J.: Proc. Natl. Acad. Sci. USA 69, 1009-1015 (1972)
Weinstein, I.B., Jeffrey, A., Jennette, K., Blobstein, S., Harvey, R., Harris, C.,
    Autrup, H., Kasai, H., Nakanishi, K.: Science 193, 592-595 (1976)
Whitmire, C.E.: J. Natl. Cancer Inst. 51, 473-478 (1973)
Whitmire, C.E., Huebner, R.J.: Science 177, 60-61 (1972)
Whitmire, C.E., Salerno, R.A.: Cancer Res. 32, 1129-1132 (1972)
Whitmire, C.E., Salerno, R.A., Rabstein, L.S., Huebner, R.J., Turner, H.C.: J. Natl.
    Cancer Inst. 47, 1255-1265 (1971)
Yang, S.K., McCourt, D.W., Roller, P.P., Gelboin, H.V.: Proc. Natl. Acad. Sci. U.S.
    73, 2594-2598 (1976)

Is Foldback DNA Repeatedly Transposed?

J. O. Bishop and C. Phillips

Foldback DNA is DNA that forms duplexes from the single-strand (denatured) state as a result of an intramolecular event. Thus, the rate of
its formation is independent of DNA concentration, unlike DNA renaturation. In principle, this can happen in two ways: (1) if a duplex
fails to separate during denaturation because of cross-linkage between
the complementary strands (other causes, such as firmly bound proteins,
would probably be considered to be artifacts), and (2) when a single
strand contains complementary sequences in the appropriate orientation.
Because of the opposite polarity of the strands of a DNA duplex, this
must always lead to the formation of a hairpin structure, with or
without a loop.

A foldback DNA molecule may contain, in addition to some duplex, a
larger or smaller amount of single-stranded DNA in the form of loops
or single-stranded tails or both. Foldback DNA has been isolated by
several different methods. These discriminate to different degrees
against molecules that contain single-stranded regions, and consequently foldback DNA prepared in different ways is not necessarily the
same.

Foldback DNA in Prokaryotes

Foldback DNA was first studied in bacteria, particularly by Doty and
his collaborators (Alberts and Doty, 1968; Alberts, 1968, Rownd et al.,
1968; Mulder and Doty, 1968). This early work was concerned with the
fact that bacterial transformation is not completely abolished by
denaturation of the transforming DNA. The residual activity was found
to be due to a class of molecules that we would now call foldback, and
it was shown that these molecules are cross-linked duplexes. Each bacterial genome contained a small number of cross-links but these were
not associated with particular genetic markers. In other words, the
cross-links occured at different positions in different copies of the
genome.

A class of foldback DNA with transforming activity was isolated and
found to constitute a very small part of the total DNA. The method
used was one that discriminates quite strongly against duplexes that
contain single-stranded regions. Simultaneously, Chevallier and Bernardi (1968) used hydroxylapatite, which discriminates in this way to
a much lesser extent, to isolate foldback DNA from *H. influenzae*. Significantly, these workers found much larger amounts of foldback DNA,
with a lower specific transforming activity.

More recently, a class of foldback sequence has been recognised in
bacteria by completely different means. These are the insertion sequences (IS), found both in the bacterial chromosome and in episomes
(reviewed by Cohen, 1976). They are frequently found in pairs in
reverse-tandem configuration, and as a result single DNA strands form

looped hairpins. The most exciting property of these sequences is that they seem to promote frequent transpositions of adjacent DNA sequences that are apparently independent of the normal recombination mechanisms of the host. IS can cause the transposition of genes without loss of activity. In other cases the insertion of a transposon into a different coding sequence inactivates it, causing, e.g., extreme polar mutations. Some IS sequences can also act as strong promoters releasing a gene from control by normal regulatory mechanisms.

Controlling Elements and Mutable Loci in Eukaryotes

The genetic effects of IS have some parallels in eukaryotes. The "controlling elements" of maize and other plants are well known (review, Fincham and Sastry, 1974). Here we will consider only the phenomena that occur at the white locus in *Drosophila melanogaster*, which were painstakingly analysed by Green (1967; 1969a, b). Briefly, an unstable mutation (w^c) was recovered following X-irradiation, which could be localised to a particular site (w^a) within the white locus. Mutation at this site is independent of recombination, is premeiotic, and occurs among the gametes of one in four w^c females and one in eight w^c males (Green, 1976). A variety of different mutational events is observed, some of which are accompanied by loss of instability. These are sometimes accompanied by large or small deletions which extend from within the white locus in either direction. They sometimes include other loci, but never the whole of the white locus, and the largest may involve perhaps 1000 - 2000 kb of DNA. Several transpositions of part of the w locus to different sites on chromosome 3 have been observed, and in at least one of these the instability of the transposed locus was similar to that of the original.

The transposition of another mutable element involving the white locus was studied in detail by Ising and Ramel (1976). The size of this element is perhaps 200 kb (0.7 map units) and 37 transposition events have been mapped to 33 sites on chromosomes 1, 2, and 3. The rate of transposition in a number of crosses averaged about 1 per 2×10^4 gametes. The rate of loss of the element from one particular position was from 10 to several hundred times greater than this in different crosses, with an average in the region of 1 per 200 gametes, which is comparable to the rate at which w^c "mutates".

Thus the genetic behaviour of these mutable alleles at the white locus is very similar to that of transposable elements in bacteria. What is totally lacking, however, is information about the structure and organisation of the DNA in these cases.

Foldback DNA in Eukaryotes

Although foldback DNA was first recognised in eukaryotes by Alberts and Doty (1968), the first significant observations were made by Davidson et al. (1973) who used hydroxylapatite (HAP) to resolve foldback and non-foldback fractions of *Xenopus laevis* DNA. They found that the amount of DNA recovered in the foldback fraction is a function of the single-strand length to which the DNA is sheared prior to fractionation. This implies the existence of a limited number of *foci* of foldback DNA formation, to each of which an increasing amount of

DNA is added when the single-strand length of the DNA is increased. These observations were extended by Perlman et al. (1976) who calcu-lated that the haploid genome of *Xenopus* contains about 10^5 foci (or 2×10^5 if the foci on complementary strands are counted separately). The same authors also eliminated cross-linkage as a major factor in the formation of foldback structures in DNA strands shorter than about 10 kb, although cross-linking is known to occur infrequently in eukar-yotes, as it does in prokaryotes (Rommelaere and Miller-Faures, 1975).

Several detailed electron-microscopic studies of the foldback DNA of higher eukaryotes have been carried out (Cech and Hearst, 1975; Deiniger and Schmid, 1976; Chamberlain et al., 1975). We have taken as a model that of Deiniger and Schmid because it seems the most complete. It deals with human DNA, and the authors have drawn a close parallel be-tween their own results and those obtained by Chamberlain et al. with *Xenopus* DNA. Briefly, Deiniger and Schmid show that most foldback mole-cules contain hairpin structures with an average stem length of 0.3 kb. About one-third of these have no observable internal single-strand loop, and the remaining two-thirds have loops averaging 1.6 kb (Fig. 1). The mean spacing of the hairpins is about 12 kb.

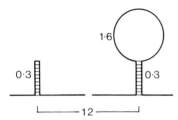

Fig. 1. Distribution of vertebrate foldback sequences according to Deiniger and Schmid (1976)

Recent experiments by Phillips (unpublished data) support this model. If *Xenopus* DNA is denatured and bound to HAP, and then eluted with a linear salt gradient, three distinct fractions are observed (Fig. 2). The first corresponds to entirely single-stranded DNA, and is almost completely sensitive to digestion by nuclease S1. The second fraction (F1-foldback) is largely but not entirely sensitive to S1-nuclease (Table 1). The S1-resistant duplexes are about 0.3 kb in length and virtually none of them can resume a double-stranded configuration if

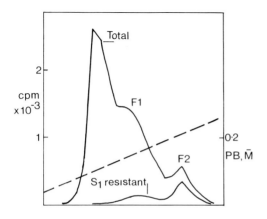

Fig. 2. Fractionation of denatured *Xenopus* DNA on HAP with a concentra-tion gradient of phosphate buffer

Table 1. Properties of foldback DNA fractions F1 and F2. DNA with a mean size of about 5 kb was fractionated as shown in Figure 2

Fraction	F1	F2
(1) Yield (% of total DNA)	25	4
(2) S1-resistance (%)	14	61
(3) Duplex yield [(1) × (2) × .01]	3.5	2.4
(4) Length of duplex (kb)	∿0.3	∿5

they are again denatured following S1 treatment. These would seem to correspond to the hairpin structures observed by electron microscopy. The third fraction (F2-foldback) is largely resistant to S1 nuclease. The S1-resistant duplexes are long and comparable in size to the single-strand length of the DNA from which it is prepared. Most of the S1-resistant material again behaves as F2-foldback if denatured after S1 treatment. The proportion of the DNA in the F2 fraction is very small when the single-strand length of the DNA is small (1 - 2% at 3 kb) and rises linearly with the single-strand length until, at 20 kb, it is comparable in quantity with the amount of F1-foldback. This fraction has properties expected of cross-linked DNA, although its cross-linked character is not proved. Its existence is the likely explanation of the apparent anomaly, noted by Cech and Hearst (1975), that the S1-sensitivity and single-strand content of foldback DNA does not increase as expected with increasing single-strand size. It also helps to explain some of the duplex structures observed in the electron microscope by Cech and Hearst (1975) and by Deiniger and Schmid (1976)..

The agreement in many points of detail between these observations and expectations based on electron-microscopic studies led us to believe that we are studying the same structures. Accordingly, we have reinterpreted the data of Perlman et al. (1976) in terms of the model shown in Figure 1.

Assuming that foldback foci are randomly distributed, the number of foci per fragment in randomly sheared DNA of a given fragment length will follow a Poisson distribution with a mean equal to the ratio of effective foci to number of fragments. The proportion of fragments that do not contain a focus is the null class of the distribution, and the remainder is the foldback fraction. We assume that the length of duplex required to bind a foldback to HAP is 0.1 kb, and that looped and un-lopped hairpins are never adjacent. The latter assumption is no doubt incorrect, but it makes little difference to the outcome while greatly simplifying the formulation. Writing the average length of a loop as L, the fragment length of the DNA as l, and the distance between adjacent stems as d, the proportion, P, of the DNA expected to fall into the foldback fraction is given by

$$P = 1 - 0.4 \exp\left\{- (l-0.2)/d\right\} - 0.6 \exp\left\{- (l-L-0.2)/d\right\} \qquad (1)$$

If the DNA is broken by means of a restriction enzyme instead of by random shear the same model is applicable as long as the cleavage sites are randomly distributed. This function is compared in Figure 3 with the data of Perlman et al. (1976) on the variation of the amount of foldback DNA with fragment length. Despite the crudity of the mathematical model the agreement is reasonably good except for very low values of l. This is to be expected since in this range the model ignores foci containing loops smaller than the average loop length of 1.6 kb.

116

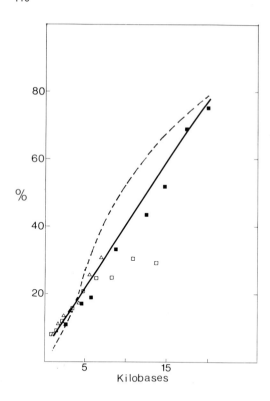

Fig. 3. Dependence of the yield of
foldback DNA on DNA fragment length,
after Perlman et al. (1976). (Δ) DNA
digested with endonuclease HindII,
and fractionated on a neutral sucrose
gradient. (□, ■) DNA broken by
digestion with nuclease S1 and fractionated on a neutral (□) or an alkaline (■) sucrose gradient. The broken
line shows the prediction of Eq. 1

Sequence Representation in Foldback DNA

Perlman et al. (1976) described two experiments that had an unexpected
outcome. As mentioned above, when the single-strand fragment length of
the DNA is short, a relatively small amount of DNA is recovered in the
foldback fraction after HAP chromatography. This contains the foldback
foci (or a proportion of them) together with relatively short contiguous DNA sequences. In such a case, if the foldback foci are located at
specific sites in the DNA, the expectation is that the foldback fraction will contain only part of the total DNA sequence.

Foldback DNA was prepared in bulk from DNA with an appropriate single-
strand length, and the proportion of the total single-copy DNA sequence
that it contained was measured. The appropriate length of about 3 kb
was obtained in two ways - by random breakage, and by cleavage with
the restriction endonuclease HindII. Some properties of the isolated
foldback DNA fractions are summarised in Table 2. There is good agreement between prediction and the amounts of DNA bound to HAP in the
tracer experiment and after the first cycle of the bulk isolation.
About half of this is lost during a second cycle of treatment for reasons which are not known.

The representation of the total DNA sequence in the foldback fraction
can be predicted on the basis of the model shown in Figure 1. Using
randomly sheared DNA, the proportion (R) of the total sequence in the
foldback fraction is approximated by

Table 2. Properties of isolated foldback DNA (Perlman et al., 1976)

Method of cleaving DNA			Endonuclease HindII	Random shear
Average length, kb			2.5[a]	3.5[b]
Predicted HAP binding, %[c]			10.4	17
Observed Binding, %	Tracer[d]		12.4	15
	Bulk	First cycle	9.4+	11.2
		Second cycle	5.3	5.6

[a] Neutral sucrose gradient
[b] Alkaline sucrose gradient
[c] From Eq. 1
[d] From Figure 3

$$R = 1 - 0.4 \exp\{-(2l-0.2)/d\} - 0.6 \exp\{-(2l-L-0.2)/d\}$$

giving, for l = 3.5 kb, R = 0.38.

The sequence representation in endonuclease-treated DNA can be better calculated as follows. The probability than an unlooped hairpin will fall between two HindII sites is $(d-0.2)(l-0.2)/d^2$ or about $(l-0.2)/d$, and for a looped hairpin it is $(d-L-0.2)(l-L-0.2)/d^2$. Hence,

$$R \sim 0.4(l-0.2)/d + 0.6 (d-L-0.2)(l-L-0.2)/d^2.$$

However, because l is not greatly different from $(\bar{L}+0.2)$, 2.5 kb as against 1.8 kb, the effect of the distribution of L is too great to be ignored. This can be dealt with crudely by assuming L to be uniformly distributed and summing $(d-L_i-0.2)(l-L_i-0.2)$ with an arbitrary class interval (0.2 kb). This gives $R \sim 0.14$. Note that in this case R = P, so that 14% is an alternative to the estimate of 10.4% shown in Table 2.

In the experiments of Perlman et al. (1976) the sequences represented in the foldback fraction were measured by driver-tracer reassociation experiments (Davidson et al., 1973). The tracer was highly labelled single-copy DNA isolated from sonicated total *Xenopus* DNA. The rate and extent to which this reassociates with unlabelled driver DNA (also sonicated) allows us to determine how much of the tracer sequence is present in the driver DNA, and whether any of it is present in a significantly higher or lower concentration than in the control. The control driver DNA is sonicated total *Xenopus* DNA. The results, which are not reproduced here, showed no difference between the control driver DNA and either of the two foldback fractions. That is to say, in each case R was apparently 1, as against expectations of 0.38 for randomly sheared DNA and 0.14 for DNA cleaved with the restriction endonuclease.

Discussion

These observations led us to propose that the foci of foldback DNA formation are present at different locations in different copies of the genome (chromosome sets). More specifically, we suggested that the foci may move from site to site during the growth of the organism in a manner analogous to transposition in bacteria. This specific suggestion now seems inherently improbable. To explain the result obtained with HindII-cleaved DNA we would have to suppose that on the average each focus occupies at least six different sites with roughly equal frequency. Since the average size of a HindII fragment is about 2.5 kb, it would then follow that an average 15 kb sequence contains on average one focus in any given genome. If the size of a coding sequence is as little as 1 kb, and if transposition is at random, the probability that a given coding sequence in a given genome contains a focus is 1/15. If, as seems likely, insertion within it inactivates a coding sequence, then any given function would be inactivated in 1/15 of genomes and in 1/225 of diploid cells. If there are 500 autonomously essential functions, two of these on average would be inactivated in every cell. On these grounds, frequent random transposition seems unlikely, and makes it worthwhile to look for other explanations of the observations. The more dramatic observation is of course that made with HindII-cleaved DNA.

One possible explanation relates to the recognition sequence of HindII, which is GT Py Pu A C (Kelly and Smith, 1970). If DNA sequences were completely random, about 35% of HindII sites would contain the internal doublet CpG. If these were partially and randomly methylated, and if this methylation prevents cleavage (the natural modification is methylation of the A residue) then the expected behaviour of HindII-cleaved DNA would tend towards that of randomly sheared DNA. However, CpG is the very doublet that is dramatically deficient in the DNA of higher eukaryotes (Subak-Sharpe et al., 1966; Russell et al., 1976), accounting for only 5% of Py-Pu doublets. When account is also taken of the frequencies of the adjacent doublets (T-Py and Pu-A) it is found that the sequence GTCGAC very likely makes up only 2% of HindII sites. Thus, explanations based on partial, random methylation of CpG sequences may be ignored.

A second type of explanation is based on polymorphism. This was brought to our attention by the work of Little (1977). In large panmictic populations, the probability that an individual carries two electrophoretic variants of a given protein is as much as 0.1. Taking into account synonymous codons, and electrophoretically neutral and balancing amino acid substitutions, Kimura (1973) pointed out that this reflects a probability of at least 10^{-3} that an individual is heterozygous in any given nucleotide pair within a coding sequence. If we allow for a rather powerful selection against changes in coding sequences, we might guess that in the remainder (more than 90%) of the genome the probability might be 10^{-2}. If so, the probability that a given HindII site is present in any two genomes is about 1 in 20. Now, the experiments of Perlman et al. (1976) were carried out with DNA from pooled livers. The number of genomes involved is not known exactly, but it was on the order of ten. On the basis of the assumptions made above, the probability that ten genomes drawn at random contain a given HindII site is less than one half. This would again tend to change the expectation from the HindII experiment towards that from random shear. However, this explanation, too, seems quite unlikely. We have already stretched the probabilities in its favour, and even as it stands, although most of the genome might be represented in the foldback fraction, we would expect the sequences to exhibit a wide range of different concentrations.

The explanation that we now favour is also based on polymorphism, but of a different sort: polymorphism, not at the nucleotide level, but at the level of transpositions. According to this hypothesis, the hairpin structures that form the foldback foci move from site to site in the genome infrequently, on an evolutionary time scale. The effects of such transpositions are most significant when they occur in the germ-line. As a result, the same focus will often occupy different positions in different genome sets, and DNA from a number of pooled genomes will give us the results that we observed. We can retain the assumption that transposition is at random, and that transposition into coding DNA inactivates its function. This would usually produce a gamete containing a recessive lethal mutation. Consequently, the transposition rate must be compatible with the spontaneous lethal mutation rate. Taking this as 2×10^{-6} per gamete per locus, and assuming that *Xenopus* has 50,000 coding sequences, gives a total recessive lethal mutation rate of 10^{-1} per gamete. If, say, half of this were due to transpositions (5×10^{-2}) and if 5% of the DNA is coding, the total transposition rate per gamete is 1. Since there are 10^5 foci, the transposition rate per gamete per focus would then be 10^{-5}. This is comparable to the transposition rates in *D. melanogaster* measured by Ising and Ramel (1976) which averaged 5×10^{-5}. From this point of view, the hypothesis seems to be tenable. It is worth pointing out here that, according to Schmid et al. (1975), hairpin structures in *D. melanogaster* DNA are about an order of magnitude larger than in the vertebrates. Thus, this particular coincidence may have no significance.

Clearly, if polymorphism of one sort or another is the basis of our observation, we would expect to obtain a different result from the DNA of a single individual. An experiment to test this is currently under way.

Acknowledgements. This work was supported by a grant to J.O.B. from the S.R.C. C.P. is a Lucky Fellow.

References

Alberts, B.: J. Mol. Biol. 32, 405-421 (1968)
Alberts, B., Doty, P.: J. Mol. Biol. 32, 379-403 (1968)
Cech, T., Hearst, J.: Cell 5, 429-446 (1975)
Chamberlain, M.E., Britten, R.J., Davidson, E.H.: J. Mol. Biol. 96, 317-333 (1975)
Chevallier, M., Bernardi, G.: J. Mol. Biol. 32, 437-452 (1968)
Cohen, S.N.: Nature (London) 263, 731-738 (1976)
Davidson, E.H., Hough, B., Amenson, C., Britten, R.J.: J. Mol. Biol. 77, 1-23 (1973)
Deiniger, P.L., Schmid, C.W.: J. Mol. Biol. 106, 773-790 (1976)
Fincham, J.R.S., Sastry, G.R.K.: Ann. Rev. Genet. 8, 15-50 (1974)
Green, M.M.: Genetics 56, 467-482 (1967)
Green, M.M.: Genetics 61, 423-428 (1969a)
Green, M.M.: Genetics 61, 429-441 (1969b)
Green, M.M.: In: The Genetics and Biology of Drosophila. Ashburner, M., Novitski, E. (eds.). New York: Academic Press, 1976 Vol. 1b, pp. 929-946
Ising, G., Ramel, C.: In: The Genetics and Biology of Drosophila. Ashburner, M., Ramel, C. (eds.). New York: Academic Press, 1976, Vol. 1b, pp. 947-954
Kelly, T.J., Smith, H.O.: J. Mol. Biol. 51, 393-409 (1970)
Kimura, M.: Cold Spring Harbor Symp. Quant. Biol. 38, 515-524 (1973)
Little, P.F.R.: Ph-D. Thesis, University of Edinburgh, 1977
Mulder, C., Doty, P.: J. Mol. Biol 32, 423-435 (1968)
Perlman, S., Phillips, C., Bishop, J.O.: Cell 8, 33-42 (1976)
Rommelaere, J., Miller-Faures, A.: J. Mol. Biol. 98, 195-218 (1975)
Rownd, R., Green, D., Sternglanz, R., Doty, P.: J. Mol. Biol. 32, 369-377 (1968)
Russell, G.J., Walker, P.M.B., Elton, R.A., Subak-Sharpe, J.H.: J. Mol. Biol. 108, 1-23 (1976)
Schmid, C., Manning, J., Davidson, N.: Cell 5, 159-172 (1975)
Subak-Sharpe, J.H., Bürk, R.R., Crawford, L.V., Morrison, J.M., Hay, J., Keir, H.M.: Cold Spring Harbor Symp. Quant. Biol. 31, 737-747 (1966)

Molecular Biology of Nitrogen Fixation

R. Dixon and C. Kennedy

Although the importance of biological nitrogen fixation in the nitrogen cycle is well established, it is recognised that a considerable increase in nitrogen input into crops will be necessary to satisfy dietary needs in future decades. There is therefore considerable interest at present in furthering our knowledge of nitrogen fixation at the molecular level. The ability to convert atmospheric nitrogen to ammonia is limited to a wide range of prokaryotic organisms, some of which can associate with higher plants to form efficient symbiotic associations. This review, however, will be primarily concerned with the free-living nitrogen-fixing bacterium *Klebsiella pneumoniae* since the biochemistry and genetics of nitrogen fixation have been extensively studied in this organism.

The nitrogen-fixing enzyme complex, termed nitrogenase, consists of two component proteins, both of which are required for activity. The larger component (Mo-Fe protein) is a tetrameric protein of molecular weight 220,000 containing molybdenum, iron, and acid-labile sulphur (Eady et al., 1972). This protein consists of two non-identical subunits distinguishable by peptide mapping and amino acid composition (Kennedy et al., 1976). The smaller component (termed Fe protein) is a dimeric protein of molecular weight 67,000 which contains iron and sulfur atoms arranged as an Fe_4S_4 cluster (Eady et al., 1972; Smith and Lang, 1974). Both proteins are extremely oxygen sensitive. Nitrogenase is a versatile enzyme in that it can reduce substrates which are stereochemically similar to dinitrogen, such as acetylene and methyl isocyanide.

Magnesium ion, ATP, anaerobic conditions, and a reductant such as sodium dithionite are required for enzyme activity in vitro. During the reaction, electrons are transferred to the Fe-protein and ATP is hydrolysed. ATP is active as $MgATP^{2-}$ and both free ATP^{4-} and $MgADP^{2-}$ are inhibitory (Thorneley and Willison, 1974; Thorneley, 1974). Electron paramagnetic resonance studies have shown that binding of $MgATP^{2-}$ to the Fe protein results in a conformational change in the molecule and a decrease in its redox potential (Smith et al., 1973). Electron transfer from the reduced $MgATP^{2-}$-Fe protein to the Mo-Fe protein allows binding of substrates and consequent substrate reduction. Evidence from Mössbauer spectroscopy with ^{57}Fe-enriched Mo-Fe protein suggests that this protein exists in a "super-reduced" form in the nitrogen-fixing system (Smith and Lang, 1974).

Acid treatment of purified Mo-Fe protein results in complete loss of activity, but releases a low-molecular-weight compound which contains both Mo and Fe. This so-called molybdenum co-factor may be a common component of many molybdenum-containing enzymes (Nason et al., 1970; 1971). Some mutants of *K. pneumoniae* produce an inactive Mo-Fe protein whose activity is restored by addition of acid-treated Mo-Fe protein prepared from wild-type organisms (St. John et al., 1975). These mutants presumably synthesise an apoprotein which lacks the molybdenum co-factor.

Genetic Analysis and Plasmid Construction

Preliminary studies using transduction with phage P1 and R factor-mediated conjugation showed that the nitrogen fixation *(nif)* genes were located close to the histidine operon on the *K. pneumoniae* chromosome (Streicher et al., 1971; Dixon and Postgate, 1971). Shanmugam et al. (1974) obtained a few deletion mutants in the *his* region of *K. pneumoniae*; some deletions extended through *his* into *nif*, others were deleted for *his*, *nif*, and *shiA* (a gene for shikimic acid permease). These results placed *nif* genes between *shiA* and the histidine operon in *Klebsiella*.

The *his nif* region has been transferred from *K. pneumoniae* to a non-nitrogen fixing bacterium *Escherichia coli* C by R-factor-mediated conjugation, resulting in the production of unstable nitrogen-fixing hybrids (Dixon and Postgate, 1972). In some hybrids the *Klebsiella* DNA was maintained by integration into the *his* region of the *E. coli* chromosome; in more unstable hybrids, covalently closed circular DNA carrying *nif* genes could be demonstrated (Cannon et al., 1974a, b). The introduction of *nif* genes into *E. coli* has facilitated the construction of F-prime factors which contain the *his nif* region and neighbouring genes (Cannon et al., 1976). The *his*, *nif* region has also been inserted into the P group of plasmid RP4 (Dixon et al., 1976). This hybrid plasmid RP41 (now termed pRD1) retains the wide host range of the P incompatability group plasmids and has allowed us to examine expression of *K. pneumoniae nif* genes in unrelated genera. When this plasmid is transferred to some *nif* mutants of *Azotobacter vinelandii*, which lack activity for Mo-Fe protein or Fe protein, the Nif[+] phenotype is restored, indicating that at least some *Klebsiella nif* genes can be expressed in this organism (Cannon and Postgate, 1976). Other non-nitrogen fixing strains such as *Agrobacterium tumefaciens* do not have a Nif[+] phenotype when carrying this plasmid, although they do produce an antigen which cross-reacts immunologically with antiserum prepared against purified *K. pneumoniae* Mo-Fe protein, suggesting that some *Klebsiella nif* DNA can be transcribed and translated in these bacteria (Dixon et al., 1976).

Although pRD1 is a useful genetic tool, it is unsuitable for in vitro experiments with *nif* DNA. Hence small amplifiable plasmids containing the *nif* region have been constructed (Cannon et al., 1977). Partial digests of pRD1 DNA were obtained by EcoR1 digestion in the presence of distamycin A, an antibiotic which binds to A-T rich sequences and protects specific sites from EcoR1 cleavage. Fragments of pRD1 were cloned on the small amplifiable plasmid pMB9 which carries the Co1E1 replicator and tetracycline resistance genes (Tc[R]). Plasmids of three different size categories were detected among His[+] Tc[R] transformants: 5.3 and 6.2 megadalton species which complemented *his*D mutations and a 14.4 megadalton plasmid which complemented *his*D and a closely linked *nif* mutation.

The *nif* Cistrons of *K. Pneumoniae*

A large number of *nif* mutants of plasmid pRD1 have been tested for genetic complementation with chromosomal *nif* derivatives of *K. pneumoniae*. Nif[-]/Nif[-] diploids were constructed by conjugal transfer of plasmid mutants into various *K. pneumoniae nif* backgrounds, and the resultant transconjugants were tested for acetylene reduction as a

measure of nitrogenase activity. Most heterogenotes contained mutations
which complemented each other to produce active nitrogenase and therefore possessed mutations in two different *nif* cistrons. Mutant plasmids
which gave background levels of nitrogenase activity were considered
to have mutations in the same *nif* cistron. The complementation pattern
obtained from 1500 distinct heterogenotes has allowed the assignment
of seven complementation groups: *nif*B, *nif*A, *nif*F, *nif*E, *nif*K, *nif*D,
and *nif*H (Kennedy and Dixon, 1977). A complex complementation pattern
is observed with some point mutants; for example, some mutants do not
complement mutations in *nif*B, *nif*A, and *nif*D. These effects are unlikely to be due to polarity because these cistrons are not contiguous
on the genetic map of *nif* (see below). They probably reflect interactions between mutant and wild-type nitrogenase components or regulatory
proteins. A few mutants were difficult to assign since they gave complementation in every combination tested. Since nitrogenase is a multi-
meric protein, it is possible that intracistronic complementation is
a common occurrence in this system; alternatively these mutants may
represent additional *nif* cistrons.

Examination of the biochemical phenotype of *nif* mutants has given an
indication of the possible functions of the various *nif* cistrons (Table
1). Extracts prepared from mutants grown under *nif* derepressing conditions have been tested for the presence of nitrogenase component
proteins on SDS polyacrylamide gels or for the presence of immunolo-
gical cross-reacting material (CRM) to antiserum prepared against
purified nitrogenase polypeptides prepared from wild type. Mutants
were also characterised with respect to activity of individual component proteins by in vitro complementation with purified wild-type
proteins.

Table 1. *Nif* cistrons of *K. pneumoniae*: mutant phenotypes

Cistron	Proteins produced by mutants		Other properties	Probable function
	Mo-Fe	Fe		
*nif*B	+	+	mutants have inactive Mo-Fe protein; activity restored by addition of acid-treated wild-type Mo-Fe protein	required for functional Mo-co-factor
*nif*A	−	−	activates *nif* derepression in trans	regulatory
*nif*F	+	+	mutant has nitrogenase activity in vitro but not in vivo	electron transport
*nif*E	variable	variable		unknown
*nif*K	variable	+	mutants have inactive Mo-Fe protein	Mo-Fe protein
*nif*D	variable	+	mutants have inactive Mo-Fe protein	Mo-Fe protein
*nif*H	+	+	mutants have inactive Fe protein	Fe protein

*Nif*B mutants produce both nitrogenase components but the Mo-Fe protein
is inactive. Activity of mutant extracts can be restored by addition
of acid-treated Mo-Fe protein prepared from the wild type, suggesting
that the *nif*B product is essential for the production of functional
molybdenum co-factor (St. John et al., 1975).

*Nif*A mutants fail to synthesise any of the nitrogenase proteins, and
therefore *nif*A is suspected to have a regulatory function. *Nif*A mutants
and *nif*B-A-F deletions are complemented by mutations in *nif*E, *nif*K,
*nif*D, and *nif*H, suggesting that the *nif*A product can activate nitrogenase synthesis in trans.

A single *nif*F mutant UN66 was isolated by St. John et al. (1975). The
mutation in this strain, *nif4066*, was transferred to plasmid pRD1 by
homogenotisation to perform complementation analysis; it complements
mutations in all other cistrons and therefore probably represents a
separate complementation group. *Nif*F mutants are apparently rare; we
have been unable to assign other mutations to this class. St. John
et al. (1975) found that UN66 had low nitrogenase activity in vivo
but high activity in vitro. Since cell-free extract assays employ
sodium dithionite as an electron donor, which by-passes the natural
electron transport pathway, they postulated that UN66 lacks a component
of this pathway.

Some *nif*E mutants produce none or very low levels of nitrogenase proteins whereas others have a full complement of both Mo-Fe and Fe proteins. We are unable to assign a function to this cistron at present.

All *nif*K and *nif*D mutants examined so far lack Mo-Fe protein activity,
but have activity for Fe protein. It is probable that these cistrons
are the structural genes for the two different sub-units of the Mo-Fe
protein (Kennedy et al., 1976). All *nif*H mutants lack Fe protein activity so this is probably the structural gene for this component.

Mapping of *nif* Mutations
===========================

Nif mutations have been ordered with respect to the histidine operon
by transductional analysis with phage P1; the frequency of Nif[+] colonies among His[+] transductants varied from 20 to 80% (Streicher et al.,
1972; St. John et al., 1975). Mutants isolated in our laboratory all
contain the same *his*D allele; the percent co-transduction frequencies
with *his*D are shown in Figure 1. An equation which relates physical
distance with co-tranduction frequencies (Wu, 1966) can be employed to
estimate the size of the *nif* region (Kennedy and Dixon, 1977). There
are apparently two *nif* gene clusters located close to *his*, separated
by a "gap" of 9 kilobases (Fig. 1). Both the *his* proximal cluster and
the *his* distal cluster span around 8 kilobases.

The precise order of *nif* mutations has been determined by 3-factor
transductional crosses (St. John et al., 1975; Kennedy and Dixon,
1977). The order indicated from many reciprocal crosses is *nif*B - *nif*A -
(*nif*L) - *nif*F - *nif*E - *nif*K - *nif*D - *nif*H. A single mutant CK265 has been
provisionally defined as *nif*L on the basis that it can be readily
separated from *nif*A mutants by transduction. It has a biochemical
phenotype similar to that of *nif*A, lacking both nitrogenase components,
and in complementation tests it behaves as a *nif*A mutant, although it
exhibits a partial trans-dominant effect on plasmid pRD1. It may represent another class of *nif*A mutant or a mutation in an undefined
cistron which exhibits a polar effect on *nif*A.

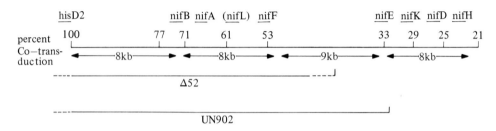

Genetic map of *nif* region in *Klebsiella pneumoniae*. The upper figures are the percent co-transduction frequencies for *nif* mutations and *his*D2. The lower figures show physical distances between markers in terms of 10^3 nucleotide base pairs (kb) derived by the Wu equation (Wu, 1966). The extent of deletions KpΔ52 and UN902 are also shown

Spontaneous deletion mutants with random end-points in *nif* genes appear to be uncommon; most spontaneous *nif* deletions lack the entire *nif* region. P2 lysogens of *K. pneumoniae* generate eductants which have deletions extending from *nif* through the *his* operon into adjacent genes (Streicher et al., 1972). Complementation and transductional analysis of a P2 eductant, KpΔ52, indicate that this strain is deleted for the proximal *nif* gene cluster, i.e., *nif*B, *nif*A, (*nif*L), and *nif*F. The precise location of the end-point of this deletion is uncertain. It must lie within the "gap" between the two clusters, raising the possibility that the attachment site for phage P2 is located within the *nif* region. Phage Mu-generated deletions of the *his-nif* region have been isolated (Bachuber et al., 1976). Strain UN906 has a *nif* deletion similar to KpΔ52, extending through *nif*B-A-(L)-F. Mapping of another Mu-generated deletion, UN902, by transduction suggested that this strain was deleted for at least *nif*B, *nif*F, and *nif*K (Bachuber et al., 1976), but complementation analysis indicates that this mutant has functional *nif*K, *nif*D, and *nif*H cistrons but is not complemented by *nif*B, *nif*A, *nif*F, or *nif*E mutants; it is therefore a *nif*B-A-F-E deletion. The contradiction between the complementation and transduction data can be interpreted if the deletion end-point lies in *nif*E, close to *nif*K, resulting in an interference in the formation of P1 transducing particles carrying *nif*K.

Functional Organisation and Regulation of *nif* Gene Clusters

The division of the *nif* region into two distinct clusters suggests that these are at least two *nif* operons, but determination of the transcriptional organisation of the seven *nif* cistrons must await polarity studies using both biochemical and genetic methods. At present we have no quantitative methods for assaying the products of the *his*-proximal *nif* cluster.

Nif genes are derepressed in the absence of ammonia. In common with other operons which determine utilization of alternative nitrogen sources such as histidine or proline, the *nif* system is subject to positive control by glutamine synthetase (GS), a protein which has both catalytic and regulatory functions (Magasanik, 1977). Evidence for glutamine synthetase control of *nif* is as follows: 1) mutations in the structural gene for glutamine synthetase (*gln*A) which abolish catalytic and regulatory properties of the enzyme, result in a failure of cells to synthesise any nitrogenase components; 2) GlnC⁻ mutants

which have derepressed levels of GS in the presence of ammonia, also
continue to synthesise nitrogenase in these conditions (Streicher et
al., 1974; Tubb, 1974). A contradiction to the preceding evidence is
that some Gln⁻ mutants produce nitrogenase in the presence of NH_4^+ even
though no glutamine synthetase protein can be detected (Shanmugam et
al., 1975). However, a *gln*A mutant of *K. aerogenes* which does not
produce catalytically active GS or CRM to the enzyme, does produce
high levels of *gln*A m-RNA, consistent with the proposal that a mutant
*gln*A product can retain regulatory properties, although it remains
undetectable by immunologic methods (Weglenski and Tyler, 1977).

Besides the general positive control mediated by GS, the *nif* genes
are undoubtedly subject to regulation by *nif*-specific products. Complementation data indicate that the *nif*A product (determined by the
his-proximal *nif* gene cluster) can activate transcription of the *his*
distal cluster. Oxygen repression of nitrogenase synthesis is probably
mediated by a *nif*-specific protein. Finally, since *nif* is subject to
repression by amino acids, even in GlnC⁻ strains, there remains the
possibility that the *nif* system is subject to negative control with
amino acids as co-repressors (Shanmugan and Morandi, 1976).

Acknowledgements. We than Ann Matthews, Linda Witts, and Eugene Kavanagh
for skilled technical assistance. We are grateful to Mike Merrick for
many useful discussions and for devising a method for homogenotisation
of *nif* mutations. Some of the mutants described here were isolated by
Adam Kondorosi and Viji Krishnapillai.

References

Bachuber, M., Brill, W.J., Howe, M.M.: J. Bacteriol. 128, 749-753 (1976)
Cannon, F.C., Dixon, R.A., Postgate, J.R.: J. Gen. Microbiol. 93, 111-125 (1976)
Cannon, F.C., Dixon, R.A., Postgate, J.R., Primrose, S.B.: J. Gen. Microbiol. 80,
    227-239 (1974a)
Cannon, F.C., Dixon, R.A., Postgate, J.R., Primrose, S.B.: J. Gen. Microbiol. 80,
    241-251 (1974b)
Cannon, F.C., Postgate, J.R.: Nature (London) 260, 271-272 (1976)
Cannon, F.C., Reidel, G.E., Ausubel, F.M.: Proc. Natl. Acad. Sci. USA 74, 2963-2967 (1977)
Dixon, R.A., Cannon, F.C., Kondorosi, A.: Nature (London) 260, 268-271 (1976)
Dixon, R.A., Postgate, J.R.: Nature (London) 234, 47-48 (1971)
Dixon, R.A., Postgate, J.R.: Nature (London) 237, 102-103 (1972)
Eady, R.R., Smith, B.E., Cook, K.A., Postgate, J.R.: Biochem. J. 128, 655-675 (1972)
Kennedy, C., Dixon, R.: In: Genetic Engineering for Nitrogen Fixation. Edited by
    A. Hollaender. New York: Plenum Publishing Corp., pp. 51-56. (1977)
Kennedy, C., Eady, R.R., Kondorosi, E., Rekosh, D.K.: Biochem. J. 155, 383-389 (1976)
Magasanik, B.: Trends in Biochem. Sci. 2, 9-12 (1977)
Nason, A., Artoine, A.D., Ketchum, P.A., Frazier III, W.A., Lee, D.K.: Proc. Natl.
    Acad. Sci. USA 65, 137-144 (1970)
Nason, A., Lee, K.-Y., Pan, S.-S., Ketchum, P.A., Lamberti, A., DeVries, J.: Proc.
    Natl. Acad. Sci. USA 68, 3242-3246 (1971)
Shanmugam, K.T., Chan, I., Morandi, C.: Biochim. Biophys. Acta. 408, 101-111 (1975)
Shanmugam, K.T., Loo, A.S., Valentine, R.C.: Biochim. Biophys. Acta. 338, 545-553
    (1974)
Shanmugam, K.T., Morandi, C.: Biochim. Biophys. Acta. 437, 322-332 (1976)
Smith, B.E., Lang, G.: Biochem. J. 137, 169-180 (1974)
Smith, B.E., Lowe, D.J., Bray, R.C.: Biochem. J. 135, 331-341 (1973)
Streicher, S.L., Gurney, E.G., Valentine, R.C.: Proc. Natl. Acad. Sci. USA 65, 74-80
    (1971)

Streicher, S.L., Gurney, E.G., Valentine, R.C.: Nature (London) <u>239</u>, 495-499 (1972)
Streicher, S.L., Shanmugam, K.T., Ausubel, F., Morandi, C., Goldberg, R.B.: J. Bacteriol. <u>120</u>, 815-821 (1974)
St. John, R.T., Johnston, M.H., Seidman, C., Garfinkel, D., Gordon, J.K., Shah, V.K., Brill, W.J.: J. Bacteriol. <u>121</u>, 759-765 (1975)
Thorneley, R.N.F.: Biochim. Biophys. Acta. <u>333</u>, 487-496 (1974)
Thorneley, R.N.F., Willison, K.R.: Biochem. J. <u>139</u>, 211-214 (1974)
Tubb, R.S.: Nature (London) <u>251</u>, 481-485 (1974)
Weglenski, P. Tyler, B.: J. Bacteriol. <u>129</u>, 880-887 (1977)
Wu, T.T.: Genetics <u>54</u>, 405-410 (1966)

On the Transfer and Expression of Prokaryotic DNA in Plant Cells Transformed by *A. tumefaciens*

J. Schell and M. van Montagu

Introduction

Crown gall is a neoplastic disease that can affect all dicotyledoneous
plants. The agent inducing this disease is a gram-negative bacterium,
Agrobacterium tumefaciens. These "crowngall" tumors are typical neo-
plasms that can proliferate autonomously in tissue cultures. Under
the action of the infecting bacteria the plant cells undergo in a few
days a neoplasmic transformation and begin to proliferate out of control. It is important to note that once the plant cells have undergone
the initial transformation, they are stably altered in their growth
properties. It has indeed been demonstrated that the tumorous character
of these crown-gall tissues is maintained in vitro in the absence of
the causative bacterium (White and Braun, 1942). For a further description of the crown-gall problem we refer to some recent reviews: Braun,
1972; Lippincott and Lippincott, 1975b; Kado, 1976.

The idea that the tumorous transformation in crown gall is the result
of a transfer of some bacterial genes to the plant cells has been around
for several years but has also been very hotly disputed. With the dis-
covery of a virulence-associated plasmid in our laboratory (Zaenen et
al., 1974) it now appears possible to settle definitely the question
of whether or not a transfer of DNA from bacteria to plants does occur
in nature.

General Properties of the Ti-Plasmids in Various *A. Tumefaciens* Strains

It is now well established that all tumor-inducing *A. tumefaciens*
strains and also *A. rhizogenes* strains contain large extrachromosomal,
covalently closed circular DNA plasmids (Zaenen et al., 1974; Schell,
1975; Watson et al., 1975; Gordon et al., 1976; Schell et al., 1976a).
It is also clear that several nononcogenic *A. radiobacter* strains also
contain large plasmids (Merlo and Nester, 1976; Schell et al., 1976b).
In some cases plasmids from oncogenic strains are more closely related
to plasmids from nononcogenic strains than to plasmids from other oncogenic strains. One can therefore assume that some of the plasmids that
are not associated with oncogenicity are mutants of the Ti-plasmids.
Certainly we have isolated several mutants of Ti-plasmids that have
lost the oncogenic determinants (see Schell et al., 1976a). On the
other hand, it has also been shown that both oncogenic and nononcogenic
Agrobacterium strains can carry large plasmids that are not associated
with oncogenicity and are probably only weakly, if at all, related to
the Ti-plasmids (Merlo and Nester, 1976; Schell et al., 1976a, b).

The plasmids from different strains have been characterized in a number
of different ways: (1) by length measurements of the circular molecules
after spreading on a water hypophase according to the Kleinschmidt
technique (Zaenen et al., 1974); (2) by comparing fingerprints of
plasmid DNA digested with restriction endonucleases (Schell et al.,

1976a; Gordon et al., 1976); (3) by direct DNA/DNA hybridization between different plasmids (Currier and Nester, 1976); (4) by electron-
microscopic studies of heteroduplex molecules formed after denaturation and reannealing of different plasmids together (Engler at al.,
1977). The results of these studies can be summarized as follows:

a) Both oncogenic and nononcogenic types of *Agrobacterium* contain
   large plasmids with sizes ranging between 95 and 156 megadaltons.

b) Several strains have been found to contain more than one of these
   large plasmids. In several cases, only one of them was associated
   with oncogenicity (the so-called Ti-plasmid).

c) The Ti-plasmids that have been studied sofar seem to fall into three
   classes:
   i) "octopine" Ti-plasmids coding for octopine (N^2-(D-1-carbo-
      xylethyl)-L-arginine) metabolism
   ii) "nopaline" Ti-plasmids coding for nopaline (N^2-(1,3-dicarbo-
      xypropyl)-L-arginine metabolism
   iii) Ti-plasmids coding for the metabolism of neither of these
      compounds

The significance of the metabolic properties of these different types
of Ti-plasmids will be discussed later.

Heteroduplex molecules made with DNA from plasmids, isolated from two
octopine strains *A. tumefaciens* B6S3 and *A. tumefaciens* ACH5, never
showed regions of nonhomology, so that we can conclude that there is
100% sequence homology. To be sure that we were not dealing with homo-
duplexes, a cointegrate plasmid between the Ti-plasmid of *A. tumefaciens*
B6S3 and the P-type plasmid RP4, was used instead of the *A. tumefaciens*
B6S3 Ti-plasmid (Schell and Van Montagu, 1977).

When hybridizations were carried out between plasmids isolated from
different nopaline-utilizing *Agrobacterium* strains, such as *Agrobacterium tumefaciens* K14 and *Agrobacterium tumefaciens* C58, different short
regions of nonhomology with a length ranging from 0.1 to 2 μm were
observed. These stretches of nonhomology were distributed over the
entire plasmid genome; many of them appeared as "eye-like" structures
suggesting that inversions of DNA sequences have occurred during plasmid evolution in *Agrobacterium*.

However, since all of these different Ti-plasmids can cause tumor formation, it was assumed that they all could have a common sequence.

We determined the homology between the two groups by hybridizing a
radioactive reference Ti-plasmid against various EcoRI/Ti-plasmid
digestions, separating them on gels, and transferring them to a nitrocellulose filter (Southern, 1975). We found only 2 EcoRI fragments
with the same electrophoretic mobility common to 5 different nopaline
Ti-plasmids and hybridizing to a radioactively labelled octopine reference Ti-plasmid.

An argument that these regions of the plasmid contain genes essential
for the expression of oncogenicity came from the analysis of a deletion
mutant in a nopaline Ti-plasmid (Hernalsteens et al., 1975). Since
this deletion mutant is unable to confer oncogenicity and has lost
exactly those two EcoRI-Ti fragments, present in most nopaline Ti-
plasmids and homologous to a segment of the octopine Ti-plasmid, we have
good indications for the correlation between this general Ti-DNA segment and the ability for tumor induction (A. Depicker et al., 1977).

Evidence for a Transfer of Genes from Ti-plasmids to Plant Cells

Genetic Evidence

That the Ti-plasmids are responsible for the oncogenic properties of *Agrobacterium* has been well established in several ways:

1. Loss of the Ti-plasmid from oncogenic strains results in the loss of oncogenicity (Van Larebeke et al., 1974; Watson et al., 1975).

2. When a Ti-plasmid is introduced into a nononcogenic acceptor strain by conjugation or transformation, this strain acquires the capacity to induce crown-gall tumors (Van Larebeke et al., 1975; Watson et al., 1975; Bomhoff et al., 1976; Schell et al., 1976a; Chilton et al., 1976; Kerr and Roberts, 1976; Kerr et al., 1977; Genetello et al., 1977; Van Larebeke et al., 1977).

3. Deletion mutants of Ti-plasmids have been isolated. *Agrobacterium* strains carrying such mutant Ti-plasmids have been shown to be non-oncogenic (Hernalsteens et al., 1975; Schell et al., 1976a).

One way to explain the association between Ti-plasmids and oncogenicity was to assume that the Ti-plasmids carry some genes that, after transfer to the plant cells, are responsible for the neoplasmic transformation of these cells.

The strongest argument in favor of this hypothesis comes from the observation that Ti-plasmids determine a new and specific synthetic property that is only observed in transformed plant cells and not in normal plant cells. Perhaps the most striking phenotypic difference between normal and crown-gall cells is the presence in the latter of the abnormal amino acids: Octopine (N^2-(D-1-carboxyethyl)-L-arginine), octopinic acid (N^2-(D-1-carboxyethyl)-L-ornithine), lysopine (N^2-(D-1-carboxyethyl)-L-lysine on the one hand and nopaline (N^2-(1,3-dicarboxy-propyl)-L-arginine on the other hand. It is mainly through the work of the group of G. Morel in Versailles that the importance of these observations was understood (see Petit et al., 1970). Indeed it was found that the type of arginine derivative synthesized in the tumor is specified by the strain of the tumor-inducing bacteria and is independent of the host plant. Although there has been some controversy about these points, they have recently been very extensively confirmed (Schilperoort and Bomhoff, 1975; Gordon et al., 1976; Kerr and Roberts, 1976). Recently, Kemp (personal communication) discovered that strains that determine octopine, octopine acid, and lysopine synthesis also specify histopine (N^2-(1-D-carboxyethyl)-L-histidine) synthesis in tumors.

A. tumefaciens strains can therefore specifically determine the biosynthesis of these "opines" in the host plant. That this capacity could be the result of a specific gene transfer from bacteria to plant cells was supported by another correlation: Those strains that induce the biosynthesis of a particular "opine" in crown-gall tissue also specifically catabolize that opine in baterial cultures. For example, "octopine strains" will induce tumors containing octopine but not nopaline and can catabolize octopine but not nopaline, and similarly "nopaline strains" induce tumors containing nopaline but not octopine and can only catabolize nopaline but not octopine (however, mutants of nopaline strains can be isolated that catabolize both nopaline and octopine - Petit and Tempé, Submitted for publ. 1977). This second correlation has now also been well documented (Petit et al., 1970; Lippincott et al., 1973; Bomhoff et al., 1976; Gordon et al., 1976; Schell et al., 1976a, b; Kerr and Roberts, 1976).

The strongest genetic argument in favor of a gene-transfer model comes from the recent demonstration that both the genes controlling opine catabolism in the bacteria and the genes determining opine synthesis in transformed plant cells, are located on the Ti-plasmid. Indeed, bacterial strains cured of the Ti-plasmid lose the capacity to catabolize opines; introduction of Ti-plasmids (via conjugation or transformation) into nononcogenic *Agrobacterium* strains unable to catabolize opines, results in the acquisition by these strains of the capacity to specifically catabolize and induce the synthesis in tumor tissues of the same opine(s) as that catabolized and induced by the donor strain of the Ti-plasmid. We were thus able to change the "nopaline strain" C58 first into a plasmid-cured nononcogenic derivative C58-C1, unable to degrade either nopaline or octopine, and subsequently into an "octopine strain" able to degrade octopine but not nopaline. This was done by introducing the Ti-plasmid from a number of "octopine strains" into the C58-C1 strain via conjugation or transformation. The tumors induced by these strains now contained octopine but no nopaline. When, on the other hand, a Ti-plasmid from a "nopaline strain" was introduced into the same C58-C1 strain, the acceptor now again degraded nopaline and induced tumors containing nopaline. Thus both the capacity to catabolize the opines and the capacity to specifically induce their synthesis in transformed plant cells are entirely determined by the type of Ti-plasmids present in these bacteria (Bomhoff et al., 1976; Schell et al., 1976a; Gordon et al., 1976; Genetello et al., 1977; Montoya et al., 1977; Van Larebeke et al., 1977). Furthermore, it has been shown that the genes determining oncogenicity and the genes controlling the opine metabolism are distinct but linked on the Ti-plasmid. This was achieved by the isolation of a series of deletion mutants of the Ti-plasmid that have lost either one or the other or both of these properties (Schell et al., 1976a, b).

Evidence from Hybridization Experiments

The search for bacterial DNA in the DNA from transformed plant cells has been going on for a long time and has yielded many controversal results and artifacts (for a review, see Kado, 1976). Recently, however, some definitive progress has been made in this field. Indeed several reports have been made by the group of Nester and Gordon in Seattle, that two fragments of the octopine Ti-plasmid, obtained after digestion with the restriction endonuclease *Sma*, specifically hybridizes with crowngall DNA and not with DNA extracted from normal plants (Chilton et al., 1977). Moreover, as was described previously, we found that octopine and nopaline Ti-plasmids only have a very limited number of Eco. R1 restriction fragments in common. One (or two?) of these fragments probably carries the genes determining oncogenicity and opine metabolism, since this fragment is absent in Ti-deletion mutants that have lost both these properties. Chilton (personal communication) has preliminary evidence indicating that it could be this fragment that hybridizes with crown-gall DNA. When both the genetic and the physical evidence is considered, a strong case can be made for the transfer of Ti-plasmid genes from *Agrobacterium* to plant cells and for the maintenance and expression of these genes in the transformed plant cells.

Mechanisms Responsible for Gene Transfer

The main question here is whether or not specific mechanisms are involved in the transfer of Ti-plasmid DNA to plant cells. If such

131

mechanisms exist one can imagine that they must operate at two stages:
(1) to enable the Ti-plasmid DNA to enter the plant cells in a specific
way (it is well known that the bacteria do not enter the plant cells)
and (2) to promote the transposition of one (or more) segment(s) of
the Ti-plasmid DNA to the plant DNA.

Evidence for a Conjugative Mechanism Involved in the Transfer

Recently, we discovered that the Ti-plasmids are conjugative plasmids,
in other words that they carry genes (*Tra* genes) that promote conjugation and transfer of Ti-plasmids between donor and acceptor bacteria
(Kerr et al., 1977; Genetello et al., 1977). The most important observation in this report is that the conjugative properties of the Ti-
plasmids are normally repressed but can be specifically induced by
octopine if they are "octopine" plasmids and probably by nopaline for
"nopaline" plasmids. Recent work in our laboratory, done in collabora-
tion with Petit and Tempé in France and A. Kerr in Australia, has shown
that both the *Tra* genes and the genes controlling opine catabolism are
probably negatively controlled by the same repressor. This conclusion
is based on the following observations: Three classes of mutants have
thus far been isolated - (1) mutants that have become constitutive
both for the *Tra* genes (they conjugate in the absence of added opines)
and for the genes controlling opine catabolism (these could be repressor mutants); (2) mutants that have become constitutive for opine
catabolism but not for conjugation (these could be mutants of an operator for opine catabolism determining genes); (3) mutants that have
become constitutive for conjugation but not for opine catabolism (these
could be mutants of an operator for *Tra* genes).

A number of observations suggest that this conjugative mechanism might
also be involved in the transfer of the Ti-plasmid from bacteria to
plant cells. Lippincott and Lippincott (1975a) have observed that bean
leaf tumors induced by *A. tumefaciens* 181 show enhanced growth in response to octopine only if this compound is present during the period
of tumor induction. Tumor induction is thermosensitive, with a critical
temperature close to 30°C (Riker, 1926). The conjugative properties of
Ti-plasmids are likewise thermosensitive, within the same range of
temperature (Tempé et al., 1977). Braun (1958) established that several
steps are involved in the tumor transformation process, only one of
which, tumor inception, is thermosensitive.

Evidence in Favor of the Idea That a Given Segment of the Ti-plasmid
is Transposible to the Plant DNA

Two preliminary lines of evidence suggest that it is a particular
segment of the Ti-plasmid that is transposed to plant DNA in crown-gall
cells:

1. Analysis of deletion mutants of the Ti-plasmid indicates that the
gene(s) specifying opine metabolism are located on a particular segment
of the Ti-plasmid (Schell et al., 1976a, b). Since crown galls induced
with wild-type Ti-plasmids systematically express opine synthesis, one
must assume that the segment carrying this gene(s) is reproducibly
transferred to the plant DNA.

2. The hybridization experiments mentioned above indicate that only
a limited segment of the Ti-plasmid hybridizes specifically with crown-
gall DNA. The question therefore arises as to what properties enable
this particular Ti-segment to be transferred to, and maintained in the
plant DNA.

We recently proposed that this segment might be a "transposon" (Schell et al., 1976a, b).

Some Ti-plasmids have been shown to form typical palindrome structures (Schell et al., 1976a). In nononcogenic mutants of the same Ti-plasmid, however, no such inverted repeats could be observed. The size of the observed inverted repeats was ± 0.32 μm and of the single-strand loop in between ± 3.55 μm.

In some instances we have observed molecules with two sets of inverted repeats with identical stem sizes (± 0.30 μm) but different single-strand loops (± 3.5 μm and 14.2 μm). However, in other oncogenic Ti-plasmids no such inverted repeats have been observed thus far.

The best argument thus far in favor of the involvement of IS elements in the oncogenic properties of the Ti-plasmid comes from observations with cointegrated plasmids consisting of a Ti-plasmid and the wide host-range P-plasmid RP4. It is well known that after mobilization of a nontransmissible plasmid by a conjugative plasmid, some acceptor bacteria contain plasmid cointegrates. These consist of one molecule of each plasmid type covalently linked to form one larger plasmid with the properties of both parental types. The recombination event leading to such cointegrates is site specific and involves IS-like elements at the recombinational locus (Kopecko and Cohen, 1975).

Cointegrated plasmids consisting of a number of different Ti-plasmids and the wide host-range conjugative P-plasmid RP4 (Datta et al., 1971) were readily obtained in our laboratory by looking for the cotransfer by conjugation of the Ti-markers for opine catabolism and the antibiotic-resistance markers of the RP4 plasmid (See Schell et al., 1976a). It is important to note here that this is the first example of a fairly stable cointegration between a "promiscuous," or wide host-range conjugative plasmid (RP4) and another, unrelated, plasmid (Ti). Furthermore, the frequency with which this cointegration occurred (10^{-3}-10^{-5}) is such that there can be no doubt that such events also take place in nature. The Ti::RP4 cointegrates that we obtained can be transferred by conjugation at a high frequency (similar to transfer frequency of RP4 alone) to several other gram-negatives (e.g., *Coli*, *Pseudomonas*, *Rhizobium*) and one can therefore assume that the cointegrate has the same wide host range as RP4 alone and that Ti-like plasmids ought not to be confined to *Agrobacterium*.

These Ti::RP4 cointegrates have been studied under the electron microscope and also by analysis of the fingerprints obtained after fragmentation with restriction-endonucleases such as EcoR1 and SacII.

All these studies have demonstrated that the Ti::RP4 cointegrates consist of a complete RP4 and a complete Ti-plasmid. The point of integration in the RP4 plasmid seems not to be random (in the Sac D fragment).

Furthermore, the following observations indicate that the recombinational event leading to the cointegration occurs between similar sequences (IS-like sequences?) in both plasmids. The majority of the cointegrates Ti.C58::RP4, when denatured and reannealed, do not show any intrastrand reannealing; however, a small fraction (1-5%) of the single-stranded molecules shows one region of intrastrand reannealing. The two single-stranded loops projecting from the double-stranded stem in these molecules have the dimensions of the RP4 (± 17 μm) and Ti plasmid (± 64 μm), respectively. The interpretation of these results is that the integration between RP4 and Ti has occurred by a recombinational event involving two similar sequences with the result that the inserted RP4 plasmid is flanked on both of its extremities by these similar sequences in the same orientation (direct repeats). In some cases one of those sequences must be "flopping" around in the DNA molecule, resulting in the RP4 DNA now being flanked by inverted repeats. The fact that some single-stranded bubbles can be seen within the double-stranded stem would indicate that the inverted repeats are not completely homologous over their entire length. These observations and conclusions, if further confirmed, are a strong indication that 1) RP4 and Ti have at least one DNA sequence in common, and 2) that this sequence has some of the properties expected from an IS element, since cointegration of the two plasmids occurs via these sequences and since it would seem that these sequences can "flop around" (invert) within the DNA molecules.

Furthermore, the cointegration event is completely reversible. Indeed these cointegrated Ti::RP4 plasmids have been shown to fall apart into the two original plasmids with a measurable frequency. The resulting plasmids have been shown to be completely identical in genetic properties to the original RP4 and Ti plasmids involved in the cointegration. The frequency with which this dissociation occurs varies somewhat but is definitely increased upon transfer of the cointegrate to *E. coli*, with the result that up to 80% of the cointegrates can be shown to have fallen apart after such a transfer. That the sequence into which RP4 integrates in the Ti-plasmid is involved in the plant-transformation mechanism is indicated by the fact that in the case of the Ti C58::RP4 cointegrate we have been able to demonstrate that *Agrobacterium* strains harboring such a cointegrate are unable to induce tumors; however, these strains recover their normal oncogenic capacity when the cointegrate falls apart to give separate Ti and RP4 plasmids. Other cointegrates such as the Ti B6S3::RP4 are normally oncogenic. This would indicate that probably Ti plasmids have more than two copies of this IS-like sequence, some of them not involved in the transposition of the oncogenic DNA.

Acknowledgements. The authors wish to thank their collaborators A. Depicker, G. Engler, C. Genetello, J.P. Hernalsteens, M. Holsters, E. Messens, B. Silva, S. Van den Elsacker, N. Van Larebeke, F. Van Vliet, N. Villarroel and I. Zaenen, for providing the information described in this article.

We also wish to thank Dr. R. Schilperoort, Dr. J. Tempé, and Dr. A. Kerr for many helpful discussions and scientific collaborations.

This work was supported by grants from the "Kankerfonds van de A.S.L.K." and from the "Fonds voor Kollektief Fundamenteel Onderzoek" (N° 10316) in Belgium.

References

Bomhoff, G., Klapwijk, P.M., Kester, H.C., Schilperoort, R.A., Hernalsteens, J.P., Schell, J.: Mol. Gen. Genet. 145, 171-181 (1976)
Braun, A.C.: Proc. Natl. Acad. Sci. USA 44, 344-349 (1958)
Braun, A.C.: Prog. Exp. Tumor Res. 15, 165-187 (1972)
Chilton, M.D., Drummond, M.H., Merlo, D.J., Sciaky, D., Montoya, A.L., Gordon, M., Nester, E.W.: Cell 11, 263-271 (1977)
Chilton, M.D., Farrand, S.K., Levin, R., Nester, E.W.: Genetics 83, 609-618 (1976)
Currier, T.C., Nester, E.W.: J. Bacteriol. 126, 157-165 (1976)

Datta, N., Hedges, R.W., Shaw, E.J., Sykes, R.P., Richmond, M.H.: J. Bacteriol. 108, 1244-1248 (1971)
Depicker, A., Van Montagu, M., Schell, J.: Biochem. Soc. Trans. 5, 931-932 (1977)
Engler, G., Van Montagu, M., Zaenen, I., Schell, J.: Biochem. Soc. Trans. 5, 930-931 (1977)
Genetello, C., Van Larebeke, N., Holsters, M., Van Montagu, M., Schell, J.: Nature (London) 265, 561-563 (1977)
Gordon, M.P., Farrand, S.K., Sciaky, D., Montoya, A., Chilton, M.D., Merlo, D., Nester, E.W.: In: Molecular Biology of Plants, Symp. Univ. Minnesota. Rubenstein, I. (ed.). New York: Academic Press, 1976, in press
Hernalsteens, J.P., Engler, G., Van Larebeke, N., Van Montagu, M., Schell, J.: Arch. Int. Physiol. Biochim. 83, 368 (1975)
Kado, C.I.: Ann. Rev. Phytopathol. 14, 265-308 (1976)
Kerr, A., Manigault, P., Tempé, J.: Nature (London) 265, 560-561 (1977)
Kerr, A., Roberts, W.P.: Physiol. Plant Pathol. 9, 205-211 (1976)
Kopecko, D.J., Cohen, S.N.: Proc. Natl. Acad. Sci. USA 72, 1373-1377 (1975)
Lippincott, B.B., Lippincott, J.A.: Plant Physiol. 56, 213-215 (1975a)
Lippincott, J.A., Beiderbeck, R., Lippincott, B.B.: J. Bacteriol. 116, 378-383 (1973)
Lippincott, J.A., Lippincott, B.B.: Ann. Rev. Microbiol. 29, 377-405 (1975b)
Merlo, D.J., Nester, E.W.: J. Bacteriol., 129, 76-80 (1977)
Montoya, A.L., Chilton, M.D., Gordon, M.P., Sciaky, D., Nester, E.W.: J. Bacteriol. 129, 101-107 (1977)
Petit, A., Delhaye, S., Tempé, J., Morel, G.: Physiol. Vég. 8, 205-213 (1970)
Ricker, A.J., J. Agric. Res., 32, 83-96 (1926)
Schell, J.: In: Genetic Manipulations with Plant Material. Ledoux, L. (ed.). New York: Plenum Press, pp. 163-181 (1975)
Schell, J., Van Montagu, M., Depicker, A., De Waele, D., Engler, G., Genetello, C., Hernalsteens, J.P., Holsters, M., Messens, E., Silva, B., Van den Elsacker, S., Van Larebeke, N., Zaenen, I.: In: Molecular Biology of Plants. Symp. Univ. Minnesota. Rubinstein, I. (ed.). New York: Academic Press, 1976a, in press
Schell, J., Van Montagu, M., Depicker, A., De Waele, D., Engler, G., Genetello, C., Hernalsteens, J.P., Holsters, M., Messens, E., Silva, B., Von den Elsacker, S., Van Larebeke, N., Zaenen, I.: In: Nucleic Acids and Protein Synthesis in Plants. Bogorad, L., Weil, J.H. (eds.). New York: Plenum Press, pp. 329-342 (1976b)
Schell, J., Van Montagu, M.: In: Conference on Genetic Engineering for Nitrogen Fixation. Hollaender, A. (ed.). New York: Brookhaven Natl. Laboratories, 1977, in 159-179 (1977)
Schilperoort, R.A., Bomhoff, G.H.: In: Genetic Manipulations with Plant Materials. Ledoux, L. (ed.). New York: Plenum Press, 1975 pp. 141-162
Southern, E.: J. Mol. Biol 98, 503-517 (1975)
Tempé, J., Petit, A., Holsters, M., Van Montagu, M., Schell, J.: Proc. Natl. Acad. Sci. USA 74, 2848-2849 (1977)
Van Larebeke, N., Engler, G., Holsters, M., Van den Elsacker, S., Zaenen, I., Schilperoort, R.A., Schell, J.: Nature (London) 252, 169-170 (1974)
Van Larebeke, N., Genetello, C., Hernalsteens, J.P., Depicker, A., Zaenen, I., Messens, E., Van Montagu, M., Schell, J.: Mol. Gen. Genet. 152, 119-124 (1977)
Van Larebeke, N., Genetello, C., Schell, J., Schilperoort, R.A., Hermans, A.K., Hernalsteens, J.P., Van Montagu, M.: Nature (London) 255, 742-743 (1975)
Watson, B., Currier, T.C., Gordon, M.P., Chilton, M.D., Nester, E.W.: J. Bacteriol. 123, 255-264 (1975)
White, P.R., Braun, A.C.: Cancer Res. 2, 597-617 (1942)
Zaenen, I., Van Larebeke, N., Teuchy, H., Van Montagu, M., Schell, J.: J. Mol. Biol. 86, 109-127 (1974)

W. Guschlbauer

Nucleic Acid Structure

An Introduction

73 figures, 9 tables. XII, 146 pages. 1976
(Heidelberg Science Library, Volume 21)
ISBN 3-540-90141-8

Contents: Methods and Techniques. –
Chemistry and Enzymology of Nucleic Acids. –
Structure and Function of DNA. – Physical
Chemistry of DNA. – The Problems of DNA
Research. – Model Systems of Nucleic Acids. –
Errors and Mutations. – The Structure of Ribo-
nucleic Acids. – Nucleic Acid. – Protein Inter-
actions.

Springer-Verlag
Berlin
Heidelberg
New York

In contrast with most books on nucleic acids,
which emphasize the enzymological aspects of
nucleic acid function, this book stresses structural aspects and their influence on function.
A large part of the book is devoted to relatively
simple model systems and to the study of
nucleosides, oligonucleotides, and polynucle-
otides. Several techniques used in the study of
nucleic acids are examined and the interactions
between nucleic acids and other compounds
(dyes and proteins, for example) and their structural implications are discussed.

Molecular Biology Biochemistry and Biophysics

Editors: A. Kleinzeller, G. F. Springer, H. G. Wittmann

Springer-Verlag
Berlin
Heidelberg
New York